10/99

What Every American Should Know About
THE MID EAST
AND OIL

James M. Day

Library of Congress Cataloging-in-Publication Data

Day, James MacDonald, 1930–
The Mid East & Oil: What Every American Should Know About
By James M. Day.
    p.   cm.
Includes bibliographical references
ISBN 0-9640104-7-X
    1. Petroleum industry and trade – Political aspects – Middle East.
2. Organization of Petroleum Exporting Countries. 3. Middle East –
History. 4. Middle East – foreign relations. 5. Arab countries – Foreign
relations. 6. Israel – Foreign relations. 7. Arab-Israeli conflict. 8.
United States – Foreign relations – Middle East. 9. Middle East –
Foreign relations – United States. I. Title

HD9576.M52D39 1998
338.2'7282'0956 —— dc21
for Library of Congress   98-35458
                                                                CIP

Published by
BRIDGER HOUSE PUBLISHERS, INC
P.O. Box 2208, Carson City, NV 89702, 1-800-729-4131
Cover design by The Right Type
Printed in the United States of America
10 9 8 7 6 5 4 3 2 1

TABLE OF CONTENTS

MAPS

FIGURES

PREFACE

What Every American Should Know About THE MID EAST & OIL is a brief modern history of the complex Middle East your high school teachers and college professors didn't tell you about. It's also about *oil* — that black gold modern civilization cannot exist without. If the Mid East didn't contain two-thirds of the world's oil, most of the nations would be wastelands dependent on foreign aid few Americans would have heard about except for the pyramids, Cleopatra, Lawrence of Arabia and the horrible bloodshed in Palestine and Lebanon. The Organization of Petroleum Exporting Countries (OPEC) wouldn't exist.

The idea for this book was born out of frustration. Our high schools and colleges do a miserable job in World History — 101. They generally end the lecture on the Mid East with the Crusades and don't tell you the nasty things the Christians did to the Muslims. After lecturing on international petroleum issues on a graduate level at a university which shall remain nameless, it hit me like a *shamal* (desert sandstorm) that most students didn't know diddly squat about Arabs other than they have lots of oil, live in the desert and one worked at the 7-Eleven. Basics, such as the Arab League wasn't minor league baseball and the Seven Sisters weren't a female rock group, had to be covered before explaining why 500,000 American troops were sent to war in 1991 in a 120 degree broiling desert without a six-pack of Coors to cool them off, the reasons those bearded guys in nightgowns aren't always friendly to Americans, or may not fill up your car with super high test if you are Jewish.

If it were not for travel, haggling with Arabs over oil deals and listening to their gripes about how they have been screwed by the West, I would not have visited the library to read dry tomes and pour through congressional hearings to confirm their rhetoric had hard facts to back them up. Also, being affiliated with oil companies for many years and employed by our government during the Arab oil embargo, I heard their accounts of events. Thus, I learned that there many versions of Mid East history.

I leave it to classical scholars to drone on about Alexander the Great, Babylonians, Assyrians and ancient potentates and civilizations except where oil is involved. The modern history of the Mid East begins with the discovery of oil a few years before World War I. That's when things started hopping — the British discovered there was something in Mesopotamia and Persia worth stealing.

I do not attempt an in-depth analysis of the peoples of the Mid East. To grasp an awareness of the evolution of Arab and Iranian cultures, particularly the Islamic religion, which greatly influences the daily lives and the role of governments in the Mid East, you must go back in time. For this, I recommend Raphael Patai's *The Arab Mind and Islam* by Caesar E. Farah. As for histories, Albert Hourani's classic, *A History of the Arab Peoples*, is dry but informative, and Bernard Lewis' excellent 1996 work, *The Middle East,* breezes through 2,000 years in 433 pages.

Hopefully, written in today's irreverent vernacular in order to avoid humorless academic sanctimoniousness, this abbreviated history will whet your appetite to read the works referenced in the footnotes and delve into the intriguing Mid East and its peoples. *This is not a college text.* Graduates from any educational level may find this primer an interesting way of filling in the blanks left by television, newspapers and the old geezer in a tweed jacket who put you to sleep during World History — 101. At the very least, readers will be able to impress that attractive person of the opposite sex with his or her sharp wit and knowledge and put down that loudmouth complaining about the price of gasoline and blaming it on "Big Oil" or the "damn Arabs." It is also a modest start in gaining an understanding what a mess the Western world made of the Mid East and why America is facing today's seemingly complex and unanswerable problems.

◆ ◆ ◆

Generally, history books are dull and written in small print that strains your eyes.[1] Only students cramming the night before an exam and professors writing dissertations wade through the morass of academic dribble. This book is to be read and enjoyed with a cold beer.

A glossary is not included. I have always found it awkward to flip back to the end of a thick tome when I ran across a foreign word or phrase for the first time and occasionally spilled my beer. Thus, I opted to explain the meaning in a footnote or sidebar. Also, Arabic words may have many meanings. For example: *Baksheesh.* In a Middle Eastern hotel, it refers to a tip or gratuity. To a beggar, it is alms or a handout. It's a way of doing business when dealing with a government official or Saudi prince — it means he wants a bribe.

Specific dates are avoided except to set time frames or place an incident or series of related events in perspective. In high school and college, I loathed memorizing dates and teachers who insisted I commit the dates of wars and the reigns of potentates to memory. For the serious student desiring to correlate the milestones in Mid Eastern history, I heartily suggest The Longman Companion to *The Middle East Since 1914* by Ritchie Ovendale as an excellent compendium of chronologies of major events with concise biographies and glossaries.

Spelling doesn't count. Historians either can't spell or can't agree on the spelling because of the difficulty in transliteration from different alphabets. Neither can cartographers. I have run across many spellings of the Saudi Arabian city Jidda, Jedda, Jeddah...or however it's spelled. Names are made confusing by their length, titles and references to ancestors. For the founder of Saudi Arabia, Abdul Aziz Ibn Abdul Rahman al-Saud, I've counted spelled or abbreviated a dozen ways. Most writers refer to the king who bore forty-three sons as Abdul Aziz or Ibn Saud. I use the latter. *Ibn* means the son of in the fashion of the Scots' "Mac," which denotes family and clan names.

---

[1] There are exceptions, such as Avi Shalaim's brief and politically astute *War and Peace in the Middle East* and Peter Mansfield's *A History of the Middle East.*

Similar to watching a football game, you need a game book describing the teams and players, particularly those who scored touchdowns or fumbled the ball. Therefore, I added short biographies in sidebars and footnotes that include trivia guaranteed to dazzle friends at your local bar. Diagrams of the plays are in the form of maps. When the British and French divided up the Ottoman Empire after World War I, most of the nations didn't exist. No one knew where Syria, Palestine, Saudi Arabia and Iraq were. Mid Eastern nations are still squabbling over who owns what — Saddam Hussein claimed Kuwait was part of Iraq as recently as 1990.

I attempted to avoid long boring chapters or breaking the text into ages, eras, or reigns of kings. Short unnumbered *chapterettes* where inserted to afford the reader frequent breaks to get another beer or hit the head and think about what he or she has just read.

A modern history of the Mid East is tainted with oil — that black goo that fouls the environment, fuels the industrial world, provides the amenities demanded by today's society, makes men and nations super-rich and has caused wars...*and will continue to cause wars.* Hence, a smattering of economic charts is necessary...but only a few...because OPEC, the international cartel organized in 1960 to control and allocate the production of *their* oil, owns 77% of the world's oil reserves.

There may be a few fusspot historians who claim I place too much emphasis on oil. My reply is that they do not mention its importance enough. The British did not grab Mesopotamia because of the shortage of sand in London; nor did the French send troops to Damascus for its exquisite damask cloth for their tables. It should come as no surprise why 20,000 men and women of America's military are stationed in the Mid East today and over 500,000 were sent to Desert Storm...It was because of oil.

Unlike many history texts, I give no credit or blame to research assistants. I also refuse to follow the practice of many authors of citing discussions with "intellectuals" — those amorphous, unemployed nameless thinkers, or "high-placed government officials" — the second deputy assistant secretaries who read secret correspondence over someone's shoulder. I will admit quoting an Arab or Iranian friend on occasion, however, to allow him to vent his spleen or recite a proverb. Thus, no one else is to blame for my irreverent opinions and conclusions. Those who disagree with me or believe I'm politically incorrect are free to read a dull history tome written by a boring scholar or write their own book.

I thank the three stalwarts who corrected my grammar and spelling, Lynne Meade, Sara Salter Ward and Winifred Pizzano. I am especially grateful to Joe Wetzel who prepared the maps from a hodgepodge of atlases.

1

BIBLICAL HISTORY – Why the Arabs Have the Oil

Scripturally, the history of Mid East oil begins in the First Book of Moses, Genesis: Chapter 15, Verse 18, when the Lord told Abraham: "Unto thy seed have I given this land, from the river of Egypt [the Nile] unto the great river, the river Euphrates." Thus, God gave the Mid East to the sons of Abraham: to Isaac, through whom the Jews claim descent, and Ishmael, from whom the Arabs profess lineage. The Bible tells us that Hagar, Ishmael's mother, was Abraham's Egyptian maid — an Arab — in one of the Bible's first admissions that there was adulterous diddling going on by the righteous.

While Moses was wandering lost in the desert for forty years, the Lord told him to take the children of Israel to the land of Canaan, "the land that shall fall unto you for an *inheritance,*" according to Numbers: Chapter 34, Verse 2. Wherefore, Isaac's Jewish children received the land God directed as their *inheritance* and Ishmael's Arab offspring ended up with the parched desert that was left over as their *inheritance,* but that had all the oil.

Christians, Jews and Muslims alike follow the teachings of Moses; hence, it is clear that God favored the Arabs with oil as part of their *inheritance.* It may also help resolve any debate whether God favored Texas over New Jersey.

Although subject to challenge, this theory is not far-fetched. It is the basis of Israel's present claim to Palestine. That is not to say that Israel doesn't have any oil. It produces a few barrels a day, the equivalent of what Saudi Arabia spills on a given afternoon. According to evangelist Jim Spillman's interpretation of Deuteronomy: Chapter 32,

Verses 8-13, there is an abundance of oil in Israel if they look in the right place.[2]

> When the Most High *divided to the nations their inheritance*...he set the bounds of the people according to the children of Israel...He found [Jacob] in a desert land, and in the waste howling wilderness; he led him about...that he might eat of the increase of the fields; and he made him to suck honey out of the rock; *and oil out of the flinty rock*.

It hasn't been that Israel hasn't tried to find oil in the land of its *inheritance.* In 1994 the Israel National Oil Company drilled a 22,000 foot dry hole near the Dead Sea at a cost of $24 million and a 15,000 foot duster in the Negev Desert that cost $16 million. Maybe the Israelis should cast aside mundane geology theories and read the Bible. An Arab theory is that Allah is punishing the Israelites for stealing Egyptian oil when they captured the Egyptian Sinai.

It has been said that the Bible led to the first oil discovery in Iraq in 1927. North of Kirkuk, large holes in the ground had been venting burning natural gas for as long as man remembers. Historians believe it is the site where Nebuchadnezzar, King of the Babylonians, tossed Shadrach, Meshach and Abed-nego into the burning fiery furnace, mentioned in Daniel: Chapter 3, Verse 27.

In far off Japan, the land of Buddhism and Shintoism, they also believe that God has his (her?) favorite peoples. The Japanese historian, Wakimura, opined the cause of Japan's defeat in 1945: "God was on the side of the nation that had the oil."

After World War I, British War Cabinet Minister Lord Curzon contended: "The Allies floated to victory on a wave of oil." The United States contributed 80% of the oil to the Allies' efforts in both World Wars. In 1991 the United States and thirty-six other nations went to war

---

[2] Jim Spillman, *The Great Treasure Hunt,* Omega (1981).

against Iraq after it invaded Kuwait and threatened Saudi Arabia in order to protect the free world's oil supply. The Saudis produced the oil for the Gulf War, proving that Allah was on their side.

QUIZ: As the United States is running out of oil, does that mean that it has squandered its inheritance and that God will not be on its side in the next war?

Answer: Read this book before answering.

QUIZ: Who are the Arabs and Jews?

Arabs: There is no single valid definition. The Second Book of Chronicles: Chapter 17, Verse 11, refers to "Arabians" who lived on the east bank of the River Jordan. This comports with anthropological theory that the Arabs originated on the Arabian Peninsula before they embarked on conquests of the Mid East and their assimilation with the diverse peoples of the region while spreading Islam. Prior to the twentieth century, Arab was a derogatory word in Syria and Egypt. The Ottoman Turks called the Bedouins *pis arablar* — "dirty arabs." Some historians define Arabs as anyone whose native language is Arabic, but that falls short, as many trace their ancestry from other civilizations. For example, some Lebanese claim lineage from Phoenicia or Armenia. Nor are all Arabs Muslim. Almost 15% of the people in the Mid East professed other religions, mainly Christian, before the twentieth century. A small percentage still do. One historian cops out by with the simple definition: anyone who claims to be an Arab. I use a broad definition in this text: Arabs are those making up the 22 member nations of the Arab League formed in 1945.

Jews: The Israelis can't agree who is a Jew. The orthodox minority claim the Reform and Conservative branches aren't strict enough in their interpretation and insist on adding the Law of Conversion to the books. There are already two laws (Law of Return and Law of Nationality) determining who is Jewish. According to Talmudic law, a Jew is anyone born of a Jewish mother or who has converted to Judaism. The Knesset added: "...and who does not belong to any other religion." I will refer to those professing to be Jews in Palestine prior to the creation of a Jewish state as Jews. After that, they're Israelis.

2

ANCIENT HISTORY — The Locals Knew Where
Their Oil Was

The ancestors of the current inhabitants of the Middle East were aware of their largess of oil 4,000 years ago. In Babylonia (the middle of Iraq to those who forgot their high school history), cuneiform tablets dating back to 2000 B.C. refer to naptu, "that which flares," which the Greeks translated into *naphtha*. The Babylonians lit up the night with the glop for Alexander the Great to impress the conqueror in 331 B.C.

Herodotus, the Greek historian, reported the use of oil in Persia and Hit, a city north of Baghdad in the fifth century B.C. Bricklayers used it for mortar and waterproofing, dubbing it *asphaltos,* hence, our English word, "asphalt." Assyrians dumped flaming *naptu* on their enemies in the seventh century B.C. Reeds dipped in oil were used for torches. However, other than for mortar, illumination or ointments, the locals found little use for it. The transportation industry as we know it was millenniums in the future. Although the automobile had yet to be invented, *naptu* was used as a cure for camel mange.

It was at Maidan-i-Nafta, "The Plain of Oil," where William Knox D'Arcy, a founder of the Anglo-Persian Oil Company (now British Petroleum or BP) is credited with the discovery of oil in Persia in 1908. The truth is the Persians knew it was there all the time. Why else would they have named the place the Plain of Oil? Typical of

Western history, nothing is really discovered until some Westerner discovers it.[3]

Also archetypal of history, D'Arcy was the fat cat who weaseled the Persian oil concession from Shah Mozaffar ed-Din Qajar in 1901. It was George Reynolds, an engineer, sweating in the 115 degree heat, desolation and dysentery of the desert wilderness described in Deuteronomy who found oil. D'Arcy stayed in England boozing it up at lavish parties with his actress wife and never went to Persia.

QUIZ: Persia is now called _____.
The nation is known for the following:
- (a)   Carpets.
- (b)   Taking American hostages.
- (c   Ayatollah Khomeini.
- (d)   Oil.
- (e)   Egomaniac Shahs.

Answer: Iran. All of the above.

---

[3] In America, Colonel Edwin L. Drake is credited with the discovery of oil in Pennsylvania in 1859; however, it is well-known that the Seneca Indians used oil for caulking their canoes and sold *Seneca Oil* to the white men as medicine in the late 1700s.

3

COLONIAL HISTORY — Before Anyone Heard of OPEC

Contrary to popular belief, the Organization of Petroleum Exporting Countries (OPEC), is *not* an Arab or Mid Eastern monopolistic oil conspiracy. Venezuela, Indonesia and Nigeria are also members. To understand OPEC's workings and its Mid Eastern flavor, it is necessary to touch lightly on the history of its members and friends.

Prior to World War I, all Mid Eastern and OPEC nations were "territories" of colonial empires except Iran, Venezuela and present day Saudi Arabia. However, at the end of the war, the territories would only change masters.

Who Owned the Middle East?

In 1914 no one knew where Iraq, Israel, Palestine, Jordan, Lebanon and Syria were, although they may have read about some of them in the Bible. There were no detailed maps or boundaries. The area was part of the 400 year old Ottoman Empire and embraced everything the Turks thought worth conquering except Iran and little enclaves the British had muscled in on. The north was labeled "Greater Syria." You couldn't find Iraq on the map because the vague area was called Mesopotamia. Palestine couldn't be found either, although towns in Texas, Illinois and Arkansas were named after it.

The mountainous west coast of the Arabian peninsula bordering the Red Sea was called The Hijaz. Its only claim to fame was it contained Islam's two Holy Cities, Mecca and Medina. The Turks allowed a cantankerous old man who claimed to be a descendant of the Prophet

Middle East Today
and
Ottoman Empire
1914

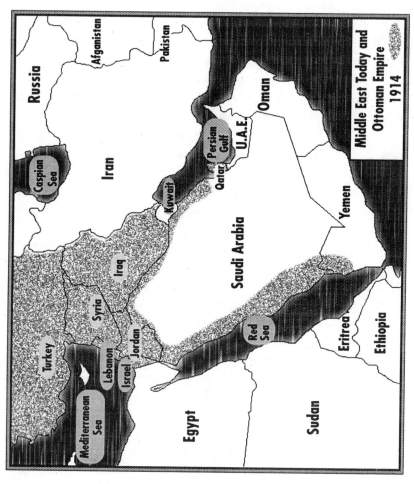

Muhammad, Sherif Hussein Ibn Ali, to run the place, so long as he did what they told him.

The east coast of the Arabian peninsula along the Persian Gulf was ruled by a bunch of sheiks. Caught between the land-grabbing Turks and Persians in the 1800s, the sheiks signed protection agreements with the British, creating the "protectorates" of Kuwait, Bahrain, Qatar and the seven Trucial States. The latter are now called the United Arab Emirates. The name *Trucial* was derived from the truce between the British and the sheiks who controlled the "Pirate Coast." Hence, the question arises: Who was protecting who?

On the southern coast of the Arabian peninsula lay the British protectorates of Aden, now the Republic of Yemen, and the Sultanate of Oman, currently an independent nation still under British protection. The only true independent was Ibn Saud, sitting in the heart of the broiling desert, who had yet to form the Kingdom of Saudi Arabia. "Independent" meant that Ibn Saud was taking money from both the British and the Turks.

QUIZ: What is a protectorate?

Answer: Black's Law Dictionary defines a protectorate: A state which has transferred the management of its more important international affairs to a stronger state." The British interpretation was: "We'll supply the troops to protect British interests in your backward country and run things because you brown chaps weren't educated at Oxford or Cambridge."

The only sovereign nation in the Mid East prior to World War I was Persia, which changed its name to Iran in 1935. Persia's claim of independence was tainted by the Anglo-Russian Entente of 1907, when Britain and Russia divided it into "spheres of influence." In other words, they were really in charge. Iranians love to tell you they are not Arabs and speak Farsi rather than Arabic. Iranians call Arabs "lizard eaters" behind their backs.

As far as the British were concerned prior to World War I, the only real difference between Persia and its Arab neighbors was that *oil had been discovered in Persia by D'Arcy's Anglo-Persian Oil Company in 1908 and it was Britain's principal source of oil.*

Who Owned North Africa?

In Africa, the OPEC member nation of Algeria was conned into believing it was part of France in 1848. Prior to then it was the home, along with Tunis, Morocco and Tripoli, of the Barbary Coast pirates, which has led some to believe that many OPEC members are still pirates. After gaining its independence in 1962, it took the name Democratic and Popular Republic of Algeria. However, it is neither democratic nor popular. The army set aside the elections after Islamic fundamentalists won in 1991 and forced the President to resign. The next President was assassinated six months later. Since, an estimated 65,000 civilians have been massacred by pro-government militias or, as the government claims, "extrajudicial executions."

Libya, the second North African OPEC nation, became an Italian colony when the Italians stole it from the Turks in 1911. Until then nothing important happened there except for battles between the Barbary Coast pirates and Stephen Decatur and the United States Marines. Everyone knows "From the halls of Montezuma to the shores of Tripoli," but in truth: the Marines never got to Tripoli. The Italians ruled Libya until after World War II when the Allies stripped Italy of its colonies. In 1951 Libya became the first nation to emerge under United Nations authority. Prior to the discovery of oil in 1958, Libya's only industry was selling scrap metal left laying around after World War II.

King Idris al-Sanusi ruled Libya from 1951 until one of modern history's greatest fruitcakes, Muammar el-Qaddafi, led a coup in 1969. The nation was renamed the Socialist People's Libyan Arab Jamahiriya. This required the invention of a new word, *Jamahiriya,* which combines the Arabic words for republic and masses. Several literati detractors joke that the name was actually derived from the Creole stew, *jambalaya,* but don't believe them.

The spelling of **Qaddafi** is as jumbled and confused as the man himself. It has been transliterated: Kadafi, Kaddafi, Qadhaafi, Qathafi, Gadhafi and Khadafy. However, the pronunciation of the last two syllables remains constant: "daffy." The root of his Bedouin family name means "one who throws."

Although not an OPEC member, Egypt is an important factor in the oil world and Mid East. Not only is it an oil exporter, it controls the Suez Canal, a vital oil tanker route. Also, two of its leaders were rabble-rousers or heroes, depending on one's viewpoint, who instigated several wars and major oil crises.

The completion of the Suez Canal by Ferdinand de Lesseps in 1869 was the biggest thing to hit Egypt since the invention of the pyramids. Unfortunately, construction costs and fraudulent bookkeeping by the French-managed canal company and European bankers bankrupted Egypt. In 1875 Queen Victoria bought 44% of the stock in the Suez Canal for the British government so its ships were guaranteed the shortcut between England and India. Not trusting the Egyptians to protect *her* waterway, the Queen sent troops to occupy Egypt in 1882.

In 1914 the Brits, concerned about the upcoming war, told Egypt it had become a protectorate. It wasn't until 1922 that Egypt gained its independence, although Britain reserved the right to protect the Suez Canal until 1956. When the Brits left, things really started to go to hell.

The most populous Arab nation, with 60 million inhabitants, Egypt formed the United Arab Republic (UAR) with Syria in 1958. Three years later, the Syrians discovered the Egyptians wanted to run everything and pulled out of the UAR. It is now called the Arab Republic of Egypt. An extremist Islamic movement doesn't make Egypt's 4 million Coptic Christians and Western tourists feel comfortable. However, Egypt and its President, Hosni Mubarak, are buddies of the United States and rank second only to Israel in receiving American foreign aid.

THE HISTORY OF EGYPT — 101½ & 102½

Ancient Egyptian history should have been covered in the tenth grade, but there are a few things they probably didn't teach you about what separates Egypt from the rest of the Arab world. King Tut and Queen Nerfertiti were probably not Arabs, but black. Cleopatra VII, the queen who kept her throne by shacking up with Julius Caesar and Mark Antony, was Greek. Egyptians claim they are Hamites and better than the other Arabs, but most people can't tell the difference.

Pharaoh Rameses the Great gained notoriety by having over 100 children and having a condom named after him. (The connection between Rameses and oil is that latex condoms are made from a synthetic petroleum product.)

Colonial Egyptian history is often marked with Napoleon's invasion of Egypt in 1798. After setting an example of stealing artifacts from the pyramids and tombs for Western museums, Napoleon left in 1802 to conquer more important places.

In 1805 an Albanian, Muhammad Ali (no relation to the former heavyweight champ), took over again for the Turks. He ran the place until 1848 when they discovered he was crazy.

It is well established, but nevertheless *untrue,* that Ferdinand de Lesseps built the first Suez Canal. The first Suez Canal was built by Pharaoh Senusret II around 2000 B.C., the old fashioned way, with 20,000 slaves. Senusret is also renown for dividing the Egyptian government into departments. To this day, the unmanageable Egyptian bureaucracy is known as the "curse of the Pharaohs."

Europe's colonial control over Africa's Mediterranean coastline extended to the Atlantic. As North Africa was full of people who spoke Arabic, Europeans and Americans claim it is part of the Mid East, even though Morocco is west of England and all the people aren't Arabs. The

British were never good at geography even when they controlled 25% of the world. During their colonial period, they considered the area now encompassing Saudi Arabia and Iraq part of their Indian Empire.

Squeezed between Algeria and Libya was the French protectorate of Tunisia, which you may recall from studying the Phoenicians, Carthaginians and the Punic Wars in the ninth grade. Tiny Morocco on the northwestern tip of Africa suffered the indignity of being divided into French and Spanish protectorates in 1912. Tunisia and Morocco became independent in 1956. Tunisia seldom gets involved in international politics for fear some bigger nation may decide to conquer it again. Neither Arab nation has much oil and were never invited to become a member of OPEC.

You rarely hear about Morocco except in old movies — Humphrey Bogart in *Casablanca* and Bing Crosby and Bob Hope in *The Road to Morocco*. The reason is Moroccans are pro-American and don't throw rocks at tourists from Peoria. Americans should like Morocco — it was the first nation to recognize the United States in 1777.

Some geographers and historians include Sudan and Somalia in the Mid East. As both countries are poor and have little oil, that's all you need to know about them unless we have to send troops to Somalia again. In 1992 American soldiers were sent to protect food being delivered by the United Nations in an attempt to feed the starving Somalians, which was being stolen by the warlords. The American troops couldn't catch the nasty warlords, who were killing and starving innocent people, and were called home after several American soldiers were killed and mutilated and it became a political embarrassment to President Clinton.

Sudanese Proverb:
After Allah made the Sudan, Allah laughed.

Venezuela and Two Hangers-on Who Shouldn't Be In OPEC

VENEZUELA won its independence from Spain in 1821. Between 1908 and 1935, the poor nation was run by General Juan Vincente Gómez, one of the most corrupt dictators in South American history. Things still aren't perfect. In 1993 President Carlos Andrés Pérez resigned to defend himself on impeachment charges arising out of corruption. Most oilmen cheered because he was the SOB who nationalized the British and American oil companies in 1976.

The first commercial oil well was drilled by Shell in 1914; however, it was the 100,000 barrel a day gusher drilled in 1922 by George Reynolds, the same Englishman who found oil in Persia in 1908, that made Venezuela the second largest oil producing nation in the world after the United States by 1928.

Venezuela was a founding member of OPEC. It ranks sixth in world oil reserves, having almost three times the oil reserves of the United States and more than any nation outside the Mid East. However, Venezuela's national interests seldom coincide with the Arab members of OPEC. When it suits them, the Venezuelans pay no attention to OPEC resolutions on pricing and production, making a sham of the organization the media calls a "cartel." Venezuela is one reason out of many that proves OPEC is not a cartel.

VENEZUELA — The First Oil Exporter

When Columbus discovered Venezuela on his third voyage in 1498, the Indians were using oil seeping from around Lake Maracaibo for lighting, medicine and caulking. In 1539, Venezuela, named "Little Venice" by early Spanish explorers, exported the first barrel of oil to Spain to treat King Charles V's gout.

NIGERIA, the most populous black African nation, gained its independence from Britain in 1960. Since then, it's had nothing but coups and civil wars, including the bloody massacre of the Christian

Ibos by the Muslim Hausas and the starvation of the Biafrans. The country is run by a pack of crooks and is a major transit point for heroin from Asia to Europe and the United States. The revenues from its vast oil exports are siphoned off by a corrupt government while the people remain destitute. Nigeria is a perennial cheater on OPEC's pricing and production quotas.

Oil was discovered in 1956, and Nigeria joined OPEC in 1971. It ranks twelfth in world oil reserves. One of the few nice things one can say about Nigeria is that its Bonny Light crude oil is excellent and makes lots of gasoline.

INDONESIA was colonized by the Dutch in 1595 and became a part of the Netherlands in 1922. Formerly the Netherlands Indies, it was commonly called the Dutch East Indies. It rebelled in an attempt to free itself of Dutch rule in 1945, but didn't obtain independence until 1956.

Oil was discovered in Indonesia in 1885. In 1890 the Royal Dutch Company was organized in the Netherlands to exploit its oil resources. Later, the company formed the core of the petroleum giant, the Royal Dutch/Shell group, which markets worldwide as Shell.

Although it ranks seventeenth in oil reserves (about one-fiftieth of Saudi Arabia's), its heavy rate of production plus a fast growing population of 200,000,000 it must feed distinguish it from other OPEC nations. At Indonesia's current rate of oil production, its reserves will be depleted in two decades unless other major oil fields are found. It is the only OPEC nation located in the Pacific. Although an OPEC member since 1962, most oil pundits predict it will withdraw from OPEC in the next decade.

OPEC once had thirteen members, an unlucky number. To understand OPEC, it is necessary to learn a little about it's former members and why they withdrew.

GABON is believed to have been originally inhabited by pygmies, which probably made it easy for France to conquer. It became part of France in 1888 and gained its autonomy in 1960.

A heavy sulfur crude was first found in 1956 by a French government-owned company. Gabon had no business being in OPEC, which it joined in 1975. It was admitted to tone down OPEC's Arab flavor after the Yom Kippur War and Arab oil embargo in 1973. Gabon's known oil reserves, ranking it thirty-fourth in the world, were too small — less than 0.2% of the total OPEC reserves. Gabon complained that OPEC's annual dues of $1.8 million were too high and withdrew when it couldn't weasel a deal for reduced dues. One of my Saudi students told me: "OPEC should never have accepted a nation whose president's name is Omar Bongo. He sounds like he should be playing drums in a rock group."

Few noticed that **ECUADOR** withdrew in 1992 after being a member since 1973. Most folks never knew it was a member. Its oil reserves are insignificant and, like Gabon, was admitted to add non-Arab members after the Yom Kippur war and Arab oil embargo.

The Ottoman Empire and the Young Turks

During the 1500s the Muslim Ottoman Empire stretched from Asia Minor through the Middle East across North Africa and into southern Europe. Following the golden age of Suleyman I The Magnificent, the empire gradually began to decline. This has been attributed to his leaving the throne to his son, Selim II The Drunk.

By the time World War I rolled around, the only lands the Ottoman Turks still held on to were in the Mid East. The largest city, Constantinople — now Istanbul — had fallen behind the other European capitals. (Istanbul straddles the continents of Europe and Asia.) It lacked paved roads and its few sewers were overflowing, in dire need of repair and reeked of you know what.

By 1908 the bankrupt Ottoman Empire had been ruled by the autocratic Sultan Abdul Hamid since 1876, who had disbanded the Parliament and suspended the constitution. Political dissent was not tolerated. Many young educated men and military officers joined secret societies as a form of self-expression and were "waiting for the day." These ranged from the Freemasons to the Committee of Union and Progress (CUP), the latter swearing an oath on the Koran and a gun. The day finally arrived after several junior army officers were suspected of treason and Sultan Hamid sent troops to round them up.

To the Sultan's surprise, the troops joined the young officers. CUP took control as the Young Turkey Party and forced the Sultan to abdicate, then restored the constitution and Parliament. In America, we call young insurgents in our political parties who rebel against the old guard "Young Turks."

The Young Turks, like Young Republicans, were full of ideals, but had no idea what they were doing. Their major concern was keeping their empire out of the hands of their grabby neighbors, Russia, Italy, Greece and Bulgaria. Between 1911 and 1914 they attempted to form alliances with Britain, France and Germany for *their* protection, but were snubbed. In desperation, they approached Russia offering an alliance and received a laughing *Nyet!*

When war came on the horizon in 1914, the Turks acted like the Three Stooges attempting the Watergate coverup. Secretly, the Turks and Germans began to discuss an alliance a second time. Negotiations with Germany were confusing because the Young Turks had all added *Pasha* to their names — Enver Pasha, Talat Pasha, Djemal Pasha, etc. — to let everyone know they were important and were running things by committee.

Germany was anxious to get its hands on two giant battleships the British were building for the Turks in England, but Winston Churchill, then First Lord of the Admiralty, snatched them before the Turkish sailors arrived to sail off with them. Things began to get hairy when a couple of German battleships slipped into Turkey for safety with British ships hot on their heels. The Turks demanded two million Turkish pounds in gold before agreeing to get involved in a war with Britain and France. To their surprise, the Germans sent the gold. The Turks had no choice but to join the war in Europe, but insisted they didn't agree to fight in their Mid East backyard. They also dragged their feet by claiming they had to stay neutral until they got their army organized. This meant that they had to tell the British that Germany had sold Turkey the two battleships trapped in their harbor. The British became skeptical when the two dreadnoughts left port flying the Turkish flag with a German crew wearing fezzes, like Shriners at a convention. The German admiral, fed up with the Young Turks dillydallying, shelled the Russian coastline after he left port.

The jig was up. Churchill ordered the British navy: "Commence hostilities at once against Turkey," without a formal declaration of war.

All the Young Turks desired was an alliance to protect their empire. Instead, they were dragged into a war which would strip them of their Mid East lands.

4

WORLD WAR I — What First Pissed Off the Arabs

The Arab Revolt and the Holy War – Nobody Showed Up

Early in the war, Djemal Pasha, the Turkish governor of Syria, which included almost every Arab corner of the Mid East, declared a *Jihad* (Holy War) to rid the Ottoman Empire of the British infidels and everyone else who didn't please him, including Arab rabble-rousers who didn't stay in line.

Meanwhile, in Cairo, Lord Kitchener, a famous British general who had whipped the Sudanese in Khartoum in 1890 and managed to eke out a victory in the Boar War against the Afrikaner farmers in 1902, came to the conclusion that he didn't have enough troops to fight the Turks if they attacked the Suez Canal. He got the jolly good idea to have the Arabs revolt against the Turks. His choice to lead the uprising was Hussein Ibn 'Ali, the Grand Sherif of Mecca, who Kitchener thought was somewhat akin to an Arab Muslim Pope.

The British civil service in India knew Sherif Hussein was an old blowhard and that Ibn Saud, with his fierce Bedouin warriors, was more experienced at raising a revolt. The bureaucrats in India pointed out that the Mid East was part of the *British Indian Empire* and Kitchener and the Army in Cairo should keep their noses out of their bailiwick. The official currency of the British Mid East wasn't Egyptian pounds, but Indian rupees. At the time, the British government of India was sitting 2,000 miles away in Simla, the summer capital, because the winter capital in Bombay was too bloomin' hot in the summer. Kitchener told the bureaucrats to mind their own business — he was a general and they were only paper shufflers. His position was desperate. He only had a scraggly army scraggly army of Egyptians, Indians and Australians to guard Egypt and the Suez Canal.

At first, Sherif Hussein was reluctant, believing it would be more profitable to remain neutral and follow the old Arab custom of taking money from both sides. Until June 1915, he was on Germany's payroll to spread nasty propaganda about the British. The Turks gave Sherif £50,000 to start a war against the British and the British slipped him £100,000 to revolt against the Turks. His choice became clear when he heard a rumor that the Turks planned to lop off his head and the British started making vague promises about making him "King of all the Arabs." Prone to exaggeration, Sherif was not entirely honest with the British when he told them hundreds of thousands would rally behind him and the cause of Arab independence.

In June 1916 Sherif fired the first symbolic shot at the Turkish garrison in Mecca to start the Arab revolt, which went bad from the beginning because no one liked the old bugger. Barely 1,000 showed up to throw off the yoke of Turkish tyranny. In order to fill out the Arab army, the British released 2,500 Ottoman Army Arab POWs. Fortunately for Sherif and the British, no one showed up for Djemal Pasha's Holy War either.

◆ ◆ ◆

T.E. Lawrence was a twenty-eight year old archeologist who looked about eighteen when he showed up in Cairo as a temporary second lieutenant translator and map maker assigned to Britain's Arab Bureau. (The lowest rank given an Oxford graduate and proper English gentleman.) Although he had been on digs around crusader castles in Palestine, he had never been to Arabia. When he accumulated a few weeks' leave, he conned the British Resident in Egypt into letting him tag along on a trip to Jeddah to meet Sherif Hussein and his son, Abdullah. From that moment, the young temporary lieutenant's life changed drastically. He would go down in history as *Lawrence of Arabia.*[4] However, Lawrence's version of the facts varies from documented history in his celebrated book, *The Seven Pillars of Wisdom:*

---

[4] Born on the wrong side of the blanket, Lawrence's birth certificate read Thomas Edward Chapman.

I believed these misfortunes of the revolt to be due mainly to faulty leadership, or rather to the lack of leadership, Arab
and English. So I went down to Arabia to see and consider its
great men. The first, the Sherif of Mecca, we knew to be too
aged. I found Abdullah too clever, Ali too clean, Zeid too cool.

Then I rode up-country to Faisel, and found in him the
leader with the necessary fire, and yet with reason to give
effect to our science. His tribesmen seem sufficient instru-
ments, and his hills to provide natural advantage.

Lawrence actually reported Faisel was an "absolute ripper," which
everyone in London knew was bloody high praise.

Lawrence and Faisel harassed the Turks from the Arabian peninsula
into what is now Syria, but not with the mythical success depicted by the
British press in England and the film, *Lawrence of Arabia.* Their only
major victory was the capture of the port of Aqaba. However, for over two
years, they kept 30,000 to 100,000 Turkish troops (depending on whose
figures you believe) defending the area and out of the real war in Europe.
It was Sir Edmund Allenby's British army that defeated the Turks.

It wasn't Lawrence's fault he didn't win the war single-handedly.
Faisel's small army was ill-trained and ill-equipped by design of the
British, who did not trust the Arabs sufficiently to give them artillery. At
times, the Arabs fought less than gallantly according to British standards; that is, not with a stiff upper lip in the face of tremendous odds.
When the Arabs attacked the Turkish garrison at Medina, they retreated
in panic from their first encounter with artillery. They later explained
that they had merely withdrawn to make coffee; thus, they are credited
with the first wartime coffee break.

LAWRENCE OF ARABIA

Lawrence claimed he was never the same after being beat-
en and raped by a couple of Turks when he was captured during the war. He admitted never having sex with women and

liked being whipped by men. Unlike the tall flamboyant Peter O'Toole who portrayed Lawrence of Arabia on the screen, he was only five-foot-six and slightly built. Lawrence, who rose from a lowly temporary lieutenant to colonel in three years, hated being a living legend. To escape fame, he joined the RAF as an enlisted man in 1922 under the name J.H. Ross, and the Royal Tank Corps in 1925 as T.E. Shaw. A few detractors claimed he toyed with fascism and was killed in a staged motorcycle accident a week before a scheduled meeting with Hitler in 1935. His handful of close friends contend he never forgave himself for his part in double-crossing the Arabs.

The First Double-Cross — The Sykes-Picot Agreement

Three months before the start of the Arab revolt, the British, French and Russians entered into the Sykes-Picot Agreement[5] to carve up the Ottoman Empire after the war. Under this secret pact, they planned to give Russia a slice of Turkey (no pun intended) and France what is now Syria and Lebanon. Britain would grab Palestine and Mesopotamia except for the northern region of Mosul, which Sir Mark Sykes gave to France because he thought the British didn't want a barren place full of Kurds. As there were no detailed maps defining the boundaries, they drew areas of general interest, *paying no attention to the peoples and religions of the area.*

The French were anxious to get their hands on Lebanon and Syria because of their alleged sphere of influence, which included French speaking Christian Maronites in Mount Lebanon who were being persecuted by the Turks and the northern interior where the Germans had built the Baghdad Railway connecting Europe and the Mid East.

---

[5] The underhanded scheme was called the Sykes-Picot-Sazanov Agreement until Russia pulled out of the war.

1916 Sykes - Picot Agreement

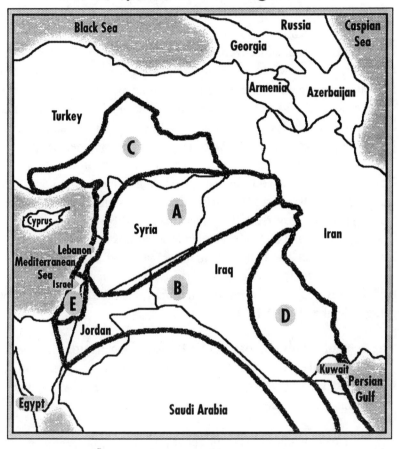

A under French influence

B under British influence

C Blue Zone, direct French control

D Red Zone, direct British control

E Allied condominium, international zone

Unfortunately, Sir Mark Sykes was an incompetent boob who would have fit into a *Monty Python* sketch.[6]   Sykes ended up on Kitchener's staff in Cairo after volunteering to go to the Mid East to "raise native scallywag corps, win over notables, or any other oddment." Sykes' lack of knowledge of the area should have been evident by his spelling of Palestine — "Filistin," but it didn't come to light until the Lords back in London discovered that he gave away the Mosul region in the north, the area where oozing naptu had been used for millenniums for waterproofing, torches and a cure for camel mange.

The area called The Hijaz, meaning "the barrier," along the Red Sea coast of the Arabian peninsula was agreed to be under British influence for the purpose of creating an Arab state. Rocky escarpments separated The Hijaz from the warring desert tribes and afforded protection for the Holy Cities of Mecca and Medina. According to local guidebooks, the Garden of Eden was located at the southern end before Adam ate the forbidden apple. However, the British had read the Bible (Genesis: Chapter 2, Verses 10-14) and knew the Garden of Eden was really some place in Mesopotamia and The Hijaz was nothing but desert and rocks, so they thought that it would be a good place to give Sherif Hussein for his Arab kingdom. Hence, it was clear that the French and the British were not going to let the old man be *King of all the Arabs.*

The Second Double-Cross – The Balfour Declaration

In November 1917 the British Foreign Secretary, Lord Balfour, wrote at letter to Lord Rothchild, Britain's leading Zionist, which was to become infamous in the Arab Mid East as the Balfour Declaration:

> I have much pleasure in conveying to you, on behalf of His Majesty's Government, the following declaration of sympathy with Jewish Zionist aspirations which have been submitted to, and approved by the Cabinet.

---

[6] If you don't believe me, read Roger Adelson's *Mark Sykes: Portrait of an Amateur.*

"His Majesty's Government views with favor the establishment in Palestine of a national home for the Jewish people, and will use its best endeavors to facilitate the achievement of this object, it being clearly understood that nothing shall be done which may prejudice the civil and religious rights of the existing non-Jewish Palestine, or the rights and political status enjoyed by Jews in any other country."

I shall be grateful if you would bring this declaration to the knowledge of the Zionist Federation.

The British were careful to call whatever the Jews might receive a *home,* rather than a state, hoping the Arabs and Jews would fall for their clever semantics. It was the Bolsheviks who let the cat out of the bag when they released a copy of the Sykes-Picot Agreement, after the revolution and Russia's withdrawal from the war, to show what a nasty imperialist Czar Nicholas II was before they shot him and the royal family.

As the Bolsheviks had no direct contact with the Arabs, they turned the copy of the Sykes-Picot Agreement over to the Turks, who were happy to pass along a copy of the dirty deal to Sherif Hussein. Sherif became suspicious of the Brits' intentions. After having himself proclaimed *King of all the Arabs,* the Brits told him he had gone a bit far and he should limit himself to being King of The Hijaz. The disadvantages of not having a telephone or fax made it necessary for Sherif to exchange correspondence regarding the size of his kingdom for two years with Sir Henry McMahon, the British High Commissioner in Egypt. In answer to Sherif's concerns, Sir Henry lied or, in the diplomatic vernacular, he was extremely tactful in his responses. Sherif was told not to worry about a Jewish *home,* after all, *it was Arab land.*

There was a certain amount of truth in McMahon's waffling. Palestine had been inhabited by the Canaanites since time immemorial, long before it was invaded by the Jews led by Joshua. And don't forget that the Philistines had been there for thousands of years and that the Israelites abandoned the area and moved to Egypt. The clinching argument was that there were over 700,000 Arabs and less than 60,000 Jews in the Holy Land, and a few more Jews wouldn't bring down property values.

But, there are two sides to the story. Sherif Hussein pussyfooted around with his revolt, insisting the British land troops first to show they were serious. Then he told the Brits they shouldn't go too far inland because it might give the local Bedouins the idea that it wasn't a real Arab revolt. Sherif was turning into a royal pain in the ass, so the British decided it was easier to promise him anything to get him off their backs so they could get on with the war. They also discovered that the old windbag couldn't produce an army, so they wouldn't have to keep their end of the deal anyway. It was too late when the British learned, as the Turkish-appointed guardian of the Holy City of Mecca, every time Sherif ran into trouble with the Bedouins he called on the Turks to send troops.

ARAB DOUBLE-DEALING
Concerned that the British wouldn't keep their promises, Faisel sent a messenger to Djemal Pasha in August 1918, warning him of British army movements and offering to switch sides if the Turks would guarantee the creation of an independent Arab nation. Lawrence's trusted second in command, Nuri al-Said, made a similar offer to a Turkish general one night while Lawrence was chasing an Arab boy around his tent. Both offers were rejected because the Turks didn't trust the Arabs and thought it was a British trick.

Why Give the Jews a Homeland?
Many people know about Theodor Herzl, who wrote *The Jewish State* in 1896, and the Zionist movement, but few are aware why a Jewish homeland became important in the middle of World War I. The answer is simple:

The British, Russians and French were getting the hell beat out of them by the Germans. The Allies needed every ally they could muster. The Italian army had failed to show up for most battles and, when they did, they were the first to surrender. Even the Turks were kicking British butts in the Dardanelles and along the Middle East frontiers. Palestine was essential for protecting the Suez Canal, Britain's shortcut to India.

Things were going so bad that the British booted out their ineffectual Prime Minister, Herbert Asquith, and installed a tough Welshman, David Lloyd George. In order to get the votes in Parliament, Lloyd George had to promise to dump Winston Churchill as First Lord of the Admiralty because he screwed up the war in Turkey and was being blamed for the horrendous defeat at Gallipoli in which the British suffered 250,000 casualties out of the 480,000 troops sent into the disaster. (Churchill fans blamed it on Kitchener who was killed on board a ship that was torpedoed early in the war.)

The French army, after suffering staggering defeats, mutinied in May 1917, resulting in their fourth — *yes, fourth* — wartime Prime Minister being tossed out on his derriére and the election of Georges Clemenceau.

Meanwhile, in Russia, Czar Nicholas II, plagued by food shortages, strikes and a mutiny by the army, abdicated in March 1917. Many of the strikes and demonstrations were fomented by a Russian Jew, Alexander Israel Helphand, who was financed by Germany.[7] On Helphand's recommendation, Germany subsidized Lenin's triumphant return to Russia by train, and it was Lenin who signed a peace treaty with Germany. Helphand had also been instrumental in arranging the alliance between the Young Turks and Germany.

As any reader of John LeCarré and Ian Fleming knows, the British recognize spies and intrigue when they see them. The Jews were not facing persecution in World War I Germany. Jews claimed they were treated better in Germany than in England. It was in Russia under the Czar where pogroms against the Jews had taken place. Rumors floated around that the Jews in Palestine planned to side with Germany. Many

---

[7] Helphand was well-named. He embezzled $25,000 from his writer-pal, Maxim Gorky, and was never trusted by Lenin and Trotsky.

of the Bolsheviks were Jews...Trotsky's real name was Bronstein. Lloyd George could see right through treachery: the Bolsheviks and Jews were conspiring to defeat the British and the French! Besides, on one liked the French.[8]

The imagined Jewish conspiracy was never uncovered. There was no Jewish uprising in Palestine. To the contrary, a Jewish Legion was formed under General Allenby's army. The Jewish Legion was pushed by Vladimir Jobotinsky, a Russian Jew, who believed that a Jewish military unit helping liberate Palestine "couldn't hurt" the chances of a Jewish homeland after the war. To make sure the Jews didn't get out of hand, the British put the Jewish Legion under the command of Lieutenant Colonel John Henry Patterson — an Irish Protestant.

What Was the United States Doing All This Time?

Uncle Sam didn't get into the war until April 2, 1917. Americans cried: "It's not our war!" President Woodrow Wilson, the son of a Presbyterian minister, didn't believe in war. Finally, when German submarines started sinking American ships and the British intercepted a secret cable from the German Foreign Secretary, Arthur Zimmermann, to Mexico seeking an alliance against the Gringos, we became sufficiently pissed and went to war.[9]

As Wilson didn't have a gripe with the Turks, the United States never declared war on the Ottoman Empire. When Wilson heard about the Sykes-Picot Agreement, he thought it was dirty pool. He was opposed to any infringement of the sovereignty of the peoples of the Ottoman Empire and in favor of their own autonomous governments. However, Wilson the do-gooder secretly approved the Balfour

---

[8] The French were still being blamed for the wave of anti-Semitism caused by the treason conviction of Captain Alfred Dreyfus, a French-Jew career soldier.

[9] The infamous Zimmermann telegram pledged, if Mexico joined Germany in the war against the United States, Germany would let Mexico have Texas, New Mexico and Arizona, which the Americans stole from Mexico in 1848.

Declaration before it was issued. He thought the Jews were only getting a "home," not a real country.

As every red-blooded American knows, once America got into the war, the Allies beat the crap out of the Germans.

There are two important facts to remember: First, the United States supplied approximately 80% of the Allies' oil. Second, the United States never went to war against the Turks or in the Mid East. It's just as well. It would have been too confusing for Americans to understand. Late in the war, the British invaded the Russian oil fields at Baku and, by the time the smoke cleared — six months after the war officially ended, the British and the Turks were allied and fighting against the Germans and the Russians in an oblivion Americans had never heard of called Turkestan.

QUIZ: What happened to the Young Turks?

Answer: Enver Pasha, Talat Pasha and Djemal Pasha ran like hell when Turkey lost. Talat fled to Germany where an Armenian blew his brains out because the Turks had killed one million Armenians during the war. Enver showed his true pinko colors by going to Russia and getting his head blown off in the Turkestan rebellion. Djemal spent the rest of his life in the pay of the Russians raising hell against the British in Afghanistan and India.

Mustafa Kemal Pasha, a Young Turk general, distanced himself from the Germans and negotiated a separate peace with the Allies. He defeated the last of the Ottoman Sultanate then drove the Greeks out of Turkey when they tried to take over. Unlike Enver and Djemal, he was a devout Muslim and an anti-communist. As Turkey's first president, he changed his name to Atatürk (Father of the Turks) so there would be no question who was in charge.

5

THE LEAGUE OF NATIONS — Churchill and the Forty Thieves

A Peace to End All Peace...or Let's Split Up the Joint[10]

If this chapterette were written in biblical vernacular, it might be phrased: "And it came to pass that the Christian tribes from Britain and France divided the Arab infidels' *inheritance* bequeathed them by the Lord after conquering the blasphemous Ottoman Empire." In truth, the French army didn't arrive until after the war.

Three things were evident to the British. First, the Russians had quit the war and weren't entitled to a slice of Turkey. Besides, the Russians didn't believe in subjugating small countries in those days. Second, the United States never declared war on Turkey and already had lots of oil, so forget Woodrow Wilson and his good intentions. They were pleased when the United States refused to join the League of Nations and could not vote against their new form of imperialism.

Third, the British had to crown a few Arab kings for the sake of appearances and assure they would still be in charge. France had its own *régime*. Since the French guillotined Louis XVI and Marie-Antoinette, they never trusted kings or queens and invented a *bureaucratie* to mismanage government. Later, France exported bureaucracy to the United States along with its bad wine the French peasants wouldn't drink.

---

[10] For a well-written account of the times, I recommend *A Peace to End All Peace: The Fall of the Ottoman Empire and the Creation of the Modern Middle East* by David Fromkin.

To understand British thinking, it must be kept in mind that Britain was being run by chaps who inherited titles and insisted on being called "Lord" or "Sir."[11] Naturally, they were astounded by the American ideals of voting and self-determination. Lord Curzon succeeded Lord Balfour as British Foreign Secretary after the war. Both were firm believers that the world was naturally divided between "governing races" and "subject races." In fact, the British have always believed that they should govern everyone. (Remember 1776?)

From the outset, it was clear the League of Nations democracies would not condone the permanent colonization of the territories of the German and Ottoman Empires. In order to placate the League members, the British and French devised "temporary mandates."[12] Mandates sounded fair and reasonable to everyone except the Arabs. However, when Lord Curzon explained the real deal to the House of Lords, it had a different ring:

> It is quite a mistake to suppose...that under the Covenant of the League...the gift of the mandate rests with the League of Nations. It does not do so. It rests with the powers who have conquered the territories, which it then falls to them to distribute.[13]

Lord Balfour let it be known who was in charge when it came to dividing up the Mid East and that he didn't give a farthing what the League of Nations, Americans and Arabs thought:

---

[11] In Britain, they still believe in knights although it's plain that chivalry died centuries ago. Winston Churchill wasn't knighted until after World War II because he was a scapegoat in Britain's disastrous campaign against the Turks at Gallipoli in 1915 and was held suspect by the English Lords because he had an American mother.

[12] The League also established mandate over other places no one ever heard of, such as Western Samoa, Togo and Tanganyika.

[13] *House of Lords Parliamentary Debates (Hansard),* 5th Series, XL col. 877 (1920).

The contradictions between the letter of the Covenant and the policy of the Allies is even more flagrant in the case of Palestine than that of the independent nation of Syria. For in Palestine we do not propose even to go through the form of consulting the wishes of the inhabitants of the country, although the American Commission has been going through the form of seeing what they were.[14]

The American Commission referred to by Lord Balfour had followed an age-old American custom of taking a poll. To everyone's surprise, the Commission found that the Syrians preferred the United States accept the mandate.[15] The British came in a poor second. Syria voted: "anybody but the French." However, public opinion didn't count much in those days, especially if you were an Arab.

President Wilson by-passed the advice of the State Department and relied on former academic colleagues to formulate his plans for the post-war world. He selected ten scholars led by, believe it or not, an expert on the Crusades. They first met in the New York Public Library before moving to the more serene surroundings of Princeton University. When they finished their report, no one noticed that it failed to mention oil.

The King-Crane Commission that Lord Balfour pooh-poohed was led by two bumblers, Henry King, the president of Oberlin College, and Charles Crane, a plumbing manufacturer from Chicago and major contributor to the Democratic Party. It is little wonder the British thought America didn't know what the hell was going on...It didn't.

---

[14] *Documents on British Foreign Policy,* Series 1, Vol. IV, pg. 345 (1952). [Memorandum dated August 19, 1919.]

[15] The Senate disdained the thought of colonialism and declined to accept a mandate over Armenia after the Turks deported the Armenians to Syria, where over one million were killed or died of starvation. No freedom-loving American would consider subjugating dark peoples. Of course, the Senate and Commission failed to mention the Virgin Islands, Hawaii, Puerto Rico and American Samoa.

Arab Kingdoms or Taking Care of the Husseins

It soon became apparent to the British that bungling Mark Sykes gave away too much to the French in the Sykes-Picot Agreement and the French hadn't done a bloody thing to help them win the war in the Mid East. With the American King-Crane Commission's report in hand, Prime Minister Lloyd George didn't mince any words about his feelings when he told the French that no one liked them and they didn't deserve Syria.

> Except for Great Britain no one had contributed anything more than a handful of black troops to the expedition in Palestine. The British now had 500,000 men on Turkish soil...and had incurred hundreds of thousands of casualties in the war against Turkey. The other governments had only put in a few nigger policemen to see that we did not steal the Holy Sepulchre![16]

The British insisted Faisel Hussein deserved to rule Syria. They showed the French newsreels of Faisel's army liberating Damascus, but failed to mention that the Turks had already pulled out and General Allenby's British army had stopped for tea in order to allow Faisel to march in triumphantly. Their argument was based on Faisel's count that his army of 100,000 had liberated Syria. The Brits were aware Faisel used Arab "romantic arithmetic" in counting his troops. Faisel's claim was also supported by the American newsman, Lowell Thomas, who was on a world lecture tour showing films of Lawrence of Arabia called *The Last Crusade.* Lawrence knew the tale was fluff, but he agreed with Faisel's story because he liked being photographed wearing a *ghutra* and *thobe* and was anxious, in his words, "to biff the French out of Syria." Also, Lawrence was aware his country had lied to the Arabs, so there was no reason why he couldn't stretch the truth a bit with the French to help the Arabs even the score.

---

[16] David Lloyd George, *War Memoirs,* Vol. 6. Boston: Little Brown (1937).

In December 1918 the British, over the nasal whining of the French, named Faisel Viceroy of Syria and told him he could call himself *Amir Faisel*. In case you're wondering what a viceroy is, it's a person ruling a country on behalf of another sovereign. The British told Faisel he was ruling on behalf of his father, Sherif Hussein, King of The Hijaz, but the Syrians knew better than to fall for that line.

"Amir" is a tricky word found in crossword puzzles, sometimes spelled "emir" or "emeer." It means commander, chieftain or prince, and was a nebulous Arab title the British were fond of bestowing when they didn't want anyone to figure out who was really in charge, including the amir. The English word "Admiral" is derived from the Arabic *amir al bahr* ("prince of the sea.")

Lawrence whisked Faisel off to the Paris Peace Conference where Faisel believed he could plead the cause of sovereignty for all the Arabs. He was wrong. Although they let him sit at the peace table, the British made it clear that he spoke only for his father's Kingdom of The Hijaz. After a year of no one listening to him, Amir Faisel returned to Damascus to take charge. He was wrong again. As soon as he started speaking about independence, the British and French got nervous.

Faisel, now wise to the Europeans' doublespeak and stranglehold of tanks and an air force that could easily hunt down Bedouins on horses and camels, wanted to go slow. However, the Syrians, who had never faced artillery, insisted on immediate independence. Faisel was caught between a sandstorm and a camel turd. Things came to a head in March 1920, when the Syrian Assembly declared the independence of Greater Syria, which included Palestine and Lebanon. As there were no boundaries, they thought Syria was just one great big place.

Meanwhile, a rump group of British bureaucrats in Mesopotamia got their signals crossed and decided Faisel's brother, Abdullah, should be king of the provinces of Basra and Baghdad.

In April 1920 the British and French set the record straight at the San Remo Conference, held on the sunny coast of Italy in order to avoid the heat of the desert and a horde of pissed off Arabs, by announcing the creation of the mandates of Syria and Lebanon under the French and Iraq and Palestine under the British.

...If you think you're confused...wait...it gets worse...

"What we want," Britain's Lord Crewe said, "is a weak and disunited Arabia, split into little principalities as far as possible under our suzerainty...incapable of co-ordinated action against us." In other words, the British wanted things muddled and confused.

After the San Remo confab, French North African Arab and Senegalese colonial troops stormed Damascus.[17] The French had to use troops from their colonies because they were afraid of another mutiny by their army. All the French soldiers were back home in France and refused go to Syria because there was no paté and escargot in the desert. The French colonial army made short work of the Syrians, then put Faisel on a train to Jerusalem and told him to find another mandate to rule.

France created Lebanon or *le Grand Libon* by taking predominately Maronite Christian Mount Lebanon they called *Petit Libon* and adding a chunk of "Greater Syria" (whatever that was), leaving Lebanon's population with a slight Christian majority in the heart of Muslim Syria.

Upon hearing that the French chucked his kid brother out of Syria, Abdullah assembled a ragtag army of 300 and started marching towards Damascus to avenge the insult to the honor of the Husseins. Fortunately for Abdullah, he travelled slowly and rested a lot. When he got to Amman, then a dusty oasis in the middle of nowhere and now the dusty capital of Jordan, Churchill and Lawrence caught up with him and told him to play it cool until made a deal with the French. Abdullah knew the British were stalling, but he wasn't anxious to face French tanks and planes, so he camped out in Amman until he got a better offer.

---

[17] The Senegalese troops were from French West Africa, now Senegal. The use of black African troops did not endear the French to the Syrians. France would eventually have to station 50,000 troops in Syria and bomb Damascus in 1925 and 1927 to maintain control.

While Abdullah was making a nuisance of himself in the desert, Faisel had taken a boat to London and was demanding some country to rule and embarrassing the British.

...Now, Churchill and the Husseins are confused...

In March 1921, Churchill, now Britain's Colonial Secretary,[18] assembled a group of bureaucrats in Cairo away from all the hullabaloo to figure out what to do with the Hussein brothers and their father, who were getting more crotchety everyday. Churchill called his advisors the "Forty Thieves" based on a tale from *Arabian Nights.* Lawrence later admitted that all the important decisions were made in London during a few dinner parties, which is probably true, because Churchill spent most of his time in the desert painting the pyramids and the Sphinx.

Unlike Ali Baba and the Forty Thieves, Winston's thieves only included two Arabs, who were brought along to avoid claims of discrimination or that they didn't listen to the Arab viewpoint. The top civil servant was Sir Percy Cox, the High Commissioner of Mesopotamia. Tagging along was his assistant, Gertrude Bell, to add a woman's touch. The knowledge of many of the Mid East "experts" appears to have been based on what they learned from the Bible in Sunday school. While it may have been progressive in London to appoint a woman, it wasn't according to the *mullahs* — the scholars and interpreters of the Koran — who had issued a *fatwa* (religious decree) declaring women were too emotional for serious deliberations: "In truth, the woman, because of her femininity, is tempted to abandon the path of reason and measure."[19]

Sir Percy's attempts to settle boundary disputes were not always successful. Mid Eastern nations are still squabbling over their borders.

---

[18] The Colonial Office was in charge of the *Wogs* — "Wily Oriental Gentleman." British geographers considered the Mid East the Orient — remember the Christmas carol; "We three Kings of Orient are...?"

[19] For an enjoyable insight into the life of women in Saudi Arabia, I suggest *The Saudis: Inside the Desert Kingdom* by Sandra Mackey.

When Sir percy couldn't reach a consensus, he had a twit and drew "neutral zones" with a red crayon, some of which still exist. Wise Ibn Saud, aware the Bedouins of Arabia wouldn't pay any attention to borders, said he was not familiar with the British system of miles. He wanted to know how long it would take a camel to walk a mile.

It soon became apparent to the Arabs that the British High Commissioners were going to run things, including telling the kings and amirs what to do. It also appeared that the heat of the desert sun had fried the brains of the Forty Thieves. Churchill's brain trust decided to name Faisel King of Mesopotamia and Abdullah Amir of *someplace*. This raised two problems.

Folks in Baghdad were expecting Abdullah to be their king and their second choice was the anti-British troublemaker, Sayid Talib. Gertrude Bell and Lady Cox solved this problem by inviting Talib to tea. When he left the tea party, Talib was tossed in the back of an armored car then put on a boat to Ceylon, a former British colony in the Indian Ocean now called Sri Lanka, and given a small pension for his time and trouble. Next month, Faisel was crowned King of Iraq. Iraq, the new name of Mesopotamia, was derived from the Arabic "well rooted." It seemed to suit the occasion. Hogs root.

Abdullah presented a bigger headache. The British and Jews didn't want an Arab running Palestine and they had run out of mandates to hand out to the Husseins. Churchill solved the enigma by drawing a few lines in the sand around Abdullah's campsite, creating a new mandate, the Amirate of Transjordan.[20]

---

[20] Transjordan was the vacant lot on the eastern side of the River Jordan. As it was across the Jordan from Palestine, it was logical they prefix the new amirate's name by adding "Trans," which means "across." The British declared it wasn't subject to the Balfour Declaration and told the Jews they couldn't buy any land in Transjordan. After Abdullah annexed the West Bank of the Jordan in 1950, he dropped the "Trans" and called his enlarged kingdom Jordan. It could be argued that its name should have reverted to Transjordan after Israel snatched it back in 1967, but it's likely that the expense of changing all the maps and road signs wasn't worth the trouble. Today the Arabs and Israel are still splitting up the West Bank.

Abdullah accepted the job as amir because he was already camped in Amman and the British promised to subsidize him as long as he didn't make any trouble. To make sure he behaved, the British put Abdullah on six month's probation. The Bedouins liked the idea that they were getting an Arab amir and thought it was their kind of place — there was nothing there but sand. The Jews complained that Transjordan was part of Palestine and should be theirs. Churchill grumbled that the Jews were lucky to get anything, then took the next boat home. In London, *The Times* described Churchill's solution as having "a disconcerting air of topsy-turvyism."

WINSTON'S HICCOUGH

Legend is that Churchill personally drew the boundary between Transjordan and Saudi Arabia with a pen and ruler after an extended liquid lunch and accidently put a squiggle in the line. Churchill's fondness for brandy is also legendary. The kink in the southern boundary came to be known as "Winston's Hiccough."

...Ibn Saud adds to the confusion...

The British had never bothered to ask Ibn Saud, who ruled next door in the little Kingdom of Nejd, if they could make Sherif Hussein King of The Hijaz. Ibn Saud had been eyeing The Hijaz for many years because one of his ancestors had his head lopped off for raiding Mecca and Medina. The Hijaz was also the font of Arabic erotic poetry, but Ibn Saud never mentioned it as a reason for wanting to conquer the place. He was still teed off over the borders drawn by Britain and the Turks for Nejd, Kuwait and Mesopotamia in 1914 he thought were none of their business. He was also still steaming from the 1922 border negotiations with the British when he lost a big chunk of Iraq, and wasn't satisfied that the Brits gave him two-thirds of its protectorate, Kuwait, to make up for it, which makes one wonder about British protection.

Ibn Saud and his Bedouin warriors were fiercely independent, and had never been wholly conquered by the Turks. The Turks claimed that there was nothing worth stealing, and pointed out that the southern part of Ibn Saud's land was called Rub al Khali or the "Empty Quarter." In fact, it was worse. That is a mistranslation of *al rabba al khali*, the "Barren Lands."

Back in 1915 when Ibn Saud was forty, Gertrude Bell was highly impressed by the tall, handsome Arab with a sweet smile and commanding appearance. To her, Ibn Saud was a consummate diplomat and polite, which was important to a proper Englishwoman and Oxford graduate. Her assessment was correct. Damn few Arabs would have condoned the unveiled woman who spoke in a shrill voice. It has been said that Ibn Saud tolerated her because she arranged his first train and automobile rides and gave him an X-ray of his hand.[21] Their first meeting was a success. Ibn Saud got guns and gold, but made no real promises.

During the war Britain paid Ibn Saud an annual subsidy of £60,000 and gave him a few rifles so he could take pot shots at the Turks in the neighborhood of Al Hasa, which Ibn Saud grabbed for his own after he ran them off. Al Hasa didn't seem important at the time because no one knew the largest oil deposits in the world were under all that sand. For his efforts, the British bestowed a knighthood on the desert warrior. Henceforth, he could call himself "Sir Abdul Aziz Ibn Saud, Knight Commander of the Most Eminent Order of the Indian Empire." Ibn Saud didn't want to be called "Sir" or have anything to do with India. He had never been out of the Arabian desert. All he wanted was guns and gold...and the rest of Arabia.

---

[21] Gertrude was no slouch. She was an astute politician and fluent in Latin, German, French, Persian, Farsi and Arabic. After her married lover was killed by the Turks at Gallipoli, she devoted her life to the Mid East and assumed the label *Umm, al-Mumminin* (Mother of the Faithful). She died from an overdose of barbiturates and was buried in Baghdad. Janet Wallach's *Desert Queen* depicts Gertrude Bell's unique and interesting life once one wades through what Gertie was wearing when she met Sheik So-and-so and the petty Victorian bureaucracy.

After the war Sherif Hussein made the serious mistake of granting himself the title "Caliph, Prince of the Faithful and Successor to the Prophet, Guardian of Islam," which would have made him the leader of Muslims throughout the world. This sent Ibn Saud and his devout Wahhabi tribesmen into a tizzy. It was one thing to lay claim to a strip of land, but it was a *kabirah,* an unpardonable sin, for the usurper to mess with their religion. Sherif's timing was off. The Brits had just cut off Ibn Saud's annual *baksheesh* of £60,000, so he had nothing to lose if he started raiding over the borders again.

Sherif's pleas to the British to defend his new Kingdom of The Hijaz were answered curtly — we stout-hearted fellows don't get involved in religious squabbles. Before Ibn Saud reached Mecca, Sherif packed his bags and ran off to Cyprus, leaving his son, Ali, in charge. Faisel and Abdullah had their hands full holding on to Iraq and Jordan and knew better than to play in Ibn Saud's sandbox. After Ibn Saud captured Medina and Jeddah, Ali turned tail and joined daddy in Cyprus.

Ibn Saud received a windfall when he captured Mecca and Medina and the rest of the west coast. It was not only an honor to be the guardian of Islam's two Holy Cities, he could charge each pilgrim a fee of £5 in gold for the *hajj* pilgrimage to the Sacred Mosque in Mecca, an obligation for every Muslim during his lifetime. Until the worldwide depression hit, about 100,000 made the pilgrimage annually.

In 1932 Ibn Saud proclaimed himself king and named the joint after his family, Saudi Arabia. The Soviet Union was the first nation to recognize his kingdom. Britain and France soon followed. The United States didn't get around to recognizing Saudi Arabia until 1939 when President Roosevelt heard that it had lots of oil.

QUIZ: Who was Sherif Hussein? Was he a crazy old coot for claiming the title "Successor of the Prophet, Guardian of Islam?"

Answer: His name was Hussein Ibn 'Ali. He was of the Hashim family, who claim to be direct descendants of the Prophet Muhammad, whose name was Muhammad Ibn Abdullah Ibn Abdul Muttalib Ibn Hashim. Every descendant of Muhammad is entitled to be called "Sherif." (This doesn't include the actor, Omar Sharif, who was born in Egypt and whose real name is Michael Shalhoub.)

Since 1073 a Hashimite had ruled Mecca as the Amir of Mecca under the title "Grand Sherif." Hence, the old king's claim can't be sneezed at. Sherif's great-grandson, King Hussein I, calls his wasteland The Hashimite Kingdom of Jordan, which may explain why he thinks he's important. Iraq's president, Saddam Hussein, is no relation, but no telling what he'll claim in the future.

Sherif abdicated and escaped to Cyprus with Standard Oil cans loaded with $10 million in gold, which proves he wasn't totally crazy.

6

PALESTINE — A British Sticky Wicket

On Paper, Everything Was Knishes and Dates

While attending the Paris Peace Conference, Faisel Hussein, acting on behalf of the Kingdom of The Hijaz, and Dr. Chaim Weizmann, a leader of the Zionist movement and later the first President of Israel, entered into an agreement recognizing the rights of the Jews to settle in Palestine. The agreement was based on their belief, or so they said, that the Arabs and Jews could live in harmony.

Historians make much ado about this nothing agreement tainted from its beginning by allegations that Faisel received money for signing it. There were also disputes as to the misleading translation made by Lawrence and Faisel's codicil containing the caveat, unless independent Arab states were established, all bets were off.

What should have raised any diplomat's or lawyer's eyebrow was the fact that Faisel did not have the authority to enter into the agreement on behalf of the Kingdom of The Hijaz, which the British had made clear was only a sliver along the western coast of the Arabian Peninsula. However, the tricky title, Amir, probably confused a lot of people. *Faisel and Weizmann parted on good terms, happy that their brethren weren't shooting at each other.*

On the Ground, Everything Was Bopkess

The biggest *fait accompli* that outraged the Zionists was Churchill's creation of Transjordan, which gave away a chunk of what they thought should a be part of Palestine to Abdullah Hussein. They were partially placated by the appointment of Sir Herbert Samuel as the first High

Commissioner of Palestine. Samuel, a Jewish Zionist, was an intelligent and fair Commissioner...perhaps too fair. He appointed anti-Zionist Arabs to key positions, including Hajj Amin Husseini, the Mufti (Islamic judge) of Jerusalem. The Mufti became the leader of the Palestinian Arabs and instituted a reign of terror he continued to direct after he was deported to Syria and later to Nazi Germany during the World War II. Vladimir Jabotinsky, who helped found the Jewish Legion during the war, organized the Jewish self-defense militia. Made up of hardened war veterans, they were more than the Arabs could handle.

During the late 1920s things became relatively tranquil, in part, because more Jews left Palestine than entered the Promised Land — a broiling hot desert to many of the *olim* (immigrants) arriving from the cold climate of Russia. Samuel's evenhanded treatment of the Arabs helped keep a lid on the rock throwing and shooting until the so-called "Wailing Wall Incident," when Jews put up a screen to separate the men and women at prayer. The Arabs objected to the takeover of the site, which is also a Muslim holy place — the Dome of the Rock and al-Aqsa mosque. Riots and terrorist attacks started again and resulted in the first of several "White Papers" issued by the British government.

The first White Paper blamed the Jewish Agency for part of the problem because of its large purchases of land and eviction of the Arab peasants, which put the British Colonial Office on the hot seat. Britain had helped establish the Jewish Agency in order to encourage Jewish immigration and land purchases. Unfortunately for the Arab peasants, the Ottoman Land Law of 1858 required that all land be registered, which most illiterate peasants ignored or were afraid to obey for fear they would be taxed or the Turks would draft them into the army. Moreover, much of the land wasn't owned by any one person, but was *ma'asha,* communal property. As in any society, there are always a few sharpies. In this case it was the Turks and Arab bigwigs who were smart enough to take advantage of the poor peasants. They registered the land as theirs and were willing to sell it, if the price was right, and the Jewish Agency had the bucks. The British limited Jewish immigration and restricted land sales for a brief period, but that didn't make either side happy. The result was more terrorism inflicted by both sides.

Hitler's rise to power and the persecution of the German Jews caused the next influx of *olim.* A second White Paper limited Jewish immigration to 15,000 a year, making it impossible to absorb the swarm of refugees escaping the holocaust. A meeting called by President Roosevelt in 1938 asking nations to accept the displaced Jews ended in failure. The United States,[22] Canada and Australia, all with wide open spaces, only offered to take a few thousand because of the worldwide Depression and unemployment, while the tiny Dominican Republic volunteered to accept up to 100,000 immigrants. Many Zionists worked against the meeting because they wanted the European Jews to go to the Promised Land. Correctly so, the Arabs thought the Western nations and Zionists were hypocrites.

The Jews were between a rock and a hard place. The Western nations were restricting their exodus from Europe; and the Germans were enslaving and killing them if they stayed. Thousands of Palestinian Jews volunteered for the British armed forces to fight against the Nazis. Others, such as the Irgun Zvai Leumi, terrorized everyone, including the British in Palestine and Egypt, where they assassinated the British Resident Minister.

Not only did the Jewish Palestinians increase their military force, the *Haganah,* during World War II, they gained political strength in America. In 1942 the Zionists met in New York City, which was home to many more Jews than Palestine, to demand that Palestine become a Jewish State after the war. Held at the Biltmore Hotel, the outcome of the meeting became known as the Biltmore Programme. The spelling may have been British, but the idea behind it was to encourage American Jews to vote for politicians who supported a Jewish state and win the public support of the Christians who deplored Hitler's atrocities.

---

[22] The United States passed immigration laws limiting the flow of all nationalities in the early 1920's. The author's father immigrated from Scotland after the law was enacted as what he jokingly called a "Canadian wetback," by taking a boat to Canada and walking across the unguarded border at Niagara Falls.

A FLAG IS BORN

A Flag is Born was a New York musical by Ben Hecht to raise tax-free funds for the Zionist cause in 1947, including the Irgun terrorists of Menachem Begin. The British filed a diplomatic protest because the Irgun was knocking off British soldiers, but gave up after Eleanor Roosevelt joined the fund-raising campaign.

7

PERSIAN HISTORY — Oil, Carpets and Nasty Shahs

You may recall your high school history teacher forcing you study the great Persian kings. The great kings were easy to remember because they were called Cyrus the Great and Darius the Great. Textbooks seldom mention Nadir Shah the Great who conquered India in 1739 (before the British thought of it) and brought back the jeweled Peacock Throne, which all later shahs loved to sit on and rule. Nadir Shah set an example for cruelty and corruption for future shahs. He was assassinated by his generals, which set an example for the Persians how to get rid of shahs.

TRIVIA QUIZ: What game is won by killing the shah?

Answer: Chess.   "Checkmate" is derived from *shah mat* meaning the "king is dead." This happens when the shah strays too far from the protection of his rook, *rukh* meaning "castle" in Persian.

After Nadir Shah, nothing important happened in Persia except for a couple of wars with Russia that gave the Czars a free hand to do anything they wanted in the northern provinces. The British took over in the south under the Anglo-Russian Entente of 1907. On paper, Persia was an independent nation and declared its neutrality at the outbreak of the war in 1914. Nevertheless, the British and Russians occupied Persia to protect Russia's southern border and the British oil fields from the Turks. The British were in a bad humor because local tribes were being paid by the Germans to dynamite their oil pipeline.

After the war the Bolsheviks relinquished the Czar's tyrannical land grabs. The British had other ideas. In 1919 they thought it would be keen if Persia became a protectorate. However, the League of Nations said that Persia didn't need protection. Undaunted, Lord Curzon offered to recognize Persia's independence, loan it £2 million and provide British military and civilian "advisors." To Curzon's chagrin, the Majlis voted the deal down because of rising nationalism and British failure to bribe enough members of the Majlis.

Majlis is the name many Mid Eastern nations call their legislature. In Saudi Arabia it is the *Majlis al Shura* — the Consultive Council. Consultive is understandable when you consider the king of Saudi Arabia doesn't have to do what the Majlis decides. It is derived from the Arabic word *yajlis,* "he sits down." Doesn't the United States Congress sit?...Or does it roost?

Broke, Persia's economy went to hell. What little wealth it had was gobbled up by the corrupt Shah Ahmad Qajar. The fact that the British government-controlled Anglo-Persian Oil Company (APOC) was cooking the books on the oil royalties didn't help the nation's treasury either.

One thing the British accomplished was to refit the Persian Cossack Brigade, 3,000 of Persia's elite calvary that were under Russian officers. After the Russian officers were called home, the British promoted Sergeant Reza Khan to colonel and put him in command. Two years later, the dashing colonel swept into Tehran with his Cossacks and declared himself prime minister. It only took Reza Khan another two years before he tossed out Shah Ahmad Qajar and conned the Majlis into declaring him Shah Reza Pahlavi, which sounded more regal.

By 1930 Persia was the fourth largest oil producer behind the United States, Russia and Venezuela. Reza Shah soon slipped into the despot's role by curtailing dissent, building an army, grabbing the best land for himself and stashing away a fortune. The first place he looked for more cash was APOC. When the oil company refused to budge, he threatened to cancel the concession, claiming the APOC concession was granted by a corrupt shah who took bribes, proving the adage: "It takes one to know one." The Brits were shocked when the Majlis unanimously upheld Reza Shah's cancellation. The Shah brought the dispute before the League of Nations, which sat on it for years until APOC and Persia came to terms and word leaked out that the League didn't have jurisdiction in commercial disputes. *Also influencing the settlement was Britain rattling its sabers by sending warships to the Persian Gulf to remind the Shah that the British government owned 51% of the stock in APOC.*

It was finally agreed to adjust the royalty by granting Persia 20% of APOC's profits derived from Persian oil and extend the term of the concession from 1961 to 1993. APOC's bookkeepers were Scotsmen from the Burmah Oil Company, APOC's second largest shareholder; thus, Reza Shah never got a fair deal. In fact, APOC never let the Persians see the books.

The British should have realized that Reza Khan had big ideas and would be difficult to handle when he changed his name to Pahlavi and established the Pahlavi dynasty, named after a language spoken by Persians centuries earlier. In 1935 the Shah changed the name of the nation to Iran, "the land of Aryans." This not only made it clear that Iranians weren't Arabs, but should have made the British suspect his pro-German leanings. Hitler's Germany was bragging that it was an Aryan nation and sending hordes of "advisors" to its cousins in Iran.

Shah Reza Pahlavi showed he was serious about the name change when he ordered the Iranian postal authorities to return all mail addressed to Persia. The Anglo-Persian Oil Company took the hint and changed its name to the Anglo-Iranian Oil Company (AIOC).

Because of the worldwide Depression and the British shortchanging Iran on its oil revenues, the nation developed a third world economy based on oil, pistachio nuts and beautiful rugs. However, they continued to sell the rugs as "Persian carpets" when they realized they could get a better price from the English, who like to decorate their mansions back home with old things.

8

MID EAST OIL — What All the Fuss Was About

By Jove, It Must Be British!

Winston Churchill is best remembered as Britain's World War II Prime Minister. An intrepid Churchill, standing like an English bulldog in the rubble of London puffing his ever present cigar and raising two fingers signifying "V for victory," was an inspiration to both the British and Americans. It is not well known that Churchill made Britain an oil power after he was named First Lord of the Admiralty in 1911 and, as Colonial Secretary after World War I, when he carved up the Mid East in Britain's best interests — *one of which was oil.*[23]

In 1912 the British became concerned about competition to APOC in the Mid East. An Armenian-born Turk, Calouste Gulbenkian, had obtained an oil concession from the Sultan of Turkey for the entire Ottoman Empire and organized the Turkish Petroleum Company. This upstart company was owned 50% by the Turkish National Bank and 25% each by the German Deutches Bank and APOC's rival, the Royal Dutch/Shell Group (Shell). Shell was split between 60% Dutch and 40% British ownership. Although British companies held a substantial interest in the Turkish National Bank, Gulbenkian owned 30% of the bank's stock. Thus the Turkish Petroleum Company wasn't sufficiently British.

---

[23] The author is not anti-British, and certainly not anti-Churchill as evidenced his naming his first bulldog "Sir Winston," whom he affectionately called "Winnie." His last bulldog was "Maggie," after the Iron Maiden, Margaret Thatcher.

The Royal Dutch/Shell Group was the result of a merger in 1901 between Royal Dutch, headed by a Dutchman, Henri Deterding, and Shell Transport & Trading, founded by an Englishman, Marcus Samuel. It was an ideal marriage although the two men never got along. Royal Dutch had extensive oil producing concessions in Indonesia and Shell was a worldwide marketer of Russian oil, whose supply was unreliable. (The Russians are still unreliable.)

Through the old boy network, a deal was cut in March 1914. Turkish Petroleum Company was reorganized, whereby APOC held 50%, the Deutches Bank and Shell 25% each, and Gulbenkian given a 5% overriding interest.

Two months after the Turkish Petroleum Company reorganization, Churchill, made a deal to purchase 51% of APOC for the British government in order to assure a supply of oil for the British navy. Britannia's navy was still ruling the waves, but needed to convert its coal-burning fleet to oil in order to reign over the German fleet.

When the vote to purchase APOC was taken on the floor of the House of Commons it sailed through 254 to 18, but the floor debate was rancorous. There were intimations that Henri Deterding, the director of Royal Dutch, was "too Dutch" and Marcus Samuel, an English Jew, "too Jewish" to be relied upon as a safe source of supply.

Britain Gets the Oil — Yankees Get Zilch

At the San Remo Conference, Britain insured it controlled the oil in its Iraq mandate. It took some shrewd negotiating, however, because Sir Mark Sykes had given away the Mosul region in the north to the French in the Sykes-Picot Agreement. However, the British had one thing going for them — their troops occupied the area.

The issue was resolved by the British confiscating the German Deutsche Bank's 25% share in the Turkish Petroleum Company and giving it to France. The French government handed the shares over to a newly established private company, Compagnie Francaise des Pétroles (CFP), and took 25% of CFP's stock as vigorish.

The Iraqi puppet government under Faisel ratified the concession of the newly formed and renamed Iraq Petroleum Company, extending its term to the year 2000. It was also decreed that the chairman must be a British subject to keep the company out of the hands of anyone who was not wearing the "old school tie."[24] The ownership of the new company was APOC 50%, Shell 25% and CFP 25%, less an overriding commission for Gulbenkian of 5%. It was a jolly good deal for the British government — it was the controlling stockholder and controlled Iraq.

America, depending on military intelligence (the CIA didn't come along until 1947), was not aware of the secret British and French agreement to divide the oil in the Mid East. The plot was uncovered by the president of Standard Oil Company of New Jersey (Exxon), who obtained a copy at the San Remo Conference and turned it over to the American ambassador. When the United States protested, Lord Curzon told the former British colony to stuff it — America had more than enough oil.

◆ ◆ ◆

The British government wasn't finished dealing. It nationalized the German Deutsche Bank's oil marketing company in Britain and handed it to APOC. Curiously, the German company was called *British Petroleum*. After Iran nationalized APOC's successor, AIOC, in 1951, it adopted the proper sounding British name, British Petroleum, then shortened it to "BP" so it could sell gasoline to unsuspecting American motorists.

---

[24] Around this time, British school and regimental ties became popular in America, but it didn't fool the British, who could immediately spot that Americans didn't have a funny accent. Two Standard Oil of New York Company (Mobil) geologists were tossed out of Iraq when they were discovered searching for "British oil."

QUIZ: What did Iraq get out of the deal?

ANSWER: Not as much as it should have. At the San Remo Conference, it was stipulated that Iraq should receive a 20% share in the concession. This was conveniently dropped in the 1925 concession agreement, and Iraq received a royalty of four shillings per ton.

Enter Uncle Sam or Open the Door

One lazy summer afternoon in 1919, the Congress began to decipher the babbling of the Department of the Interior's Bureau of Mines: *America was running out of oil!* The U.S. Geological Survey predicted a "gasoline famine" and that America would run out of oil in precisely nine years and three months. At the State Department, our diplomats in striped pants woke up to the fact that our former allies were not merely playing in a big sandbox, but had split up the Mid East into two oil concessions and hadn't invited America to bring its pail and scoop some of it up.[25]

What finally stuck in the Congress' craw was, after the United States had supplied Britain with oil during the war, the Brits wouldn't share oil that wasn't even theirs. The snotty remarks of Sir E. Mackay Edgar bragging that wherever Americans turn they will find British enterprise and capital there before them were read on the floor of the Senate:

> ...America one of these days — and not very distant days, either — will have to come to us for oil, copper, and perhaps, the iron ore she needs.[26]

---

[25] The State Department was also aware that Shell was hogging the best oil lands in Dutch controlled Indonesia and that Burmah* Oil (the largest private stockholder in APOC) had a 99 year concession in Burma and excluded Americans. (*Not a spelling error. The British spell things funny.)

[26] 58 Cong. Rec. 4785 (1919); *see also 4136.*

The threat caused Senator Phalen of California to ask why Britain hadn't paid its wartime debt to the United States of $4 billion instead of investing capital in oil exploration and maintaining troops in the Mid East to keep Americans out. He called England "the great cormorant of the world," which was about as nasty a thing you could say on the Senate floor in those days. What really teed off Senator Phalen was that the "British octopus," Shell, was drilling for oil in California. As a result, Congress passed a law denying mineral leases on federal lands to foreign corporations whose countries denied Americans the reciprocal right to mine for minerals.[27]

America demanded an "Open Door" policy to allow American oil and mining companies equal access to the mineral lands of the world barred by Britain and France. On the floor of Congress, there was talk of embargoing oil exports to Britain. At America's urging, several South American nations passed laws banning oil concessions to state-owned oil companies such as APOC and CFP. Because of the anti-British sentiment in Congress, Britain decided to open the door...but only a crack.

Seven American oil companies were interested in taking a piece of the Iraq Petroleum Company, but the British haggled so much, five said the hell with it. Britain would not discuss giving up anything in Persia — it was all theirs and most Americans either didn't know where it was or believed it was too far away. American oilmen also knew the country wasn't running out of oil, as the bureaucrats predicted, because they were finding gushers in Oklahoma and Texas.[28]

---

[27] The Mineral Leasing Act of 1920. The law gave the Secretary of the Interior the discretion whether or not to lease public lands in the United States, which encompassed almost 40% of the total land in America. Today, the federal government owns approximately 30% of the nation's lands and is the largest recipient of oil and gas royalties in the country — several billion dollars annually.

[28] During World War I, Interior Department bureaucrats reported that no more oil was to be found in Texas.

American interests were led by 300 pound Walter Teagle, the president of Standard Oil of New Jersey (Exxon). Everyone knew he was in charge because he was called "The Boss." Teagle was well-suited for the job. He had worked in his father's Cleveland refinery and fought Rockefeller's Standard Oil Company Trust's voracious gobbling up of 85% of America's refinery capacity. When Rockefeller finally squeezed his father's company dry and brought it into the Rockefeller empire, he hired the son because young Walter was the kind of man Rockefeller liked: big, tough and smart.

After six years of wrangling, The Boss negotiated a 23.75% interest in the Iraq Petroleum Company for Exxon and Mobil to split. This left 23.75% each for Shell, APOC and CFP and 5% for Gulbenkian. Teagle knew that 11.875% was all he was going to get for Exxon by July 1928 because oil had recently been discovered in Iraq's Kurdish region of Mosul where Nebuchadnezzar had thrown Shadrach, Meshach and Abed-nego into the fiery furnace. It also helps explain why the Kurds never got their own nation during the big carve up of Iraq and Syria after the war.

KURDISH HISTORY — 101 1/2

The Kurds are a distinct ethnic people living in Iraq, Iran, Armenia, Turkey and Syria. In the 1920 Peace Treaty of Sèvres, between the Ottoman Empire and the Allies, the Kurds were granted an independent homeland in Anatolia (Turkey) and the province of Mosul, which was never ratified by the League of Nations. Instead, "Iraq was created by Churchill, who had the mad idea of joining two widely separated oil wells, [at] Kirkuk and Mosul, by uniting three widely separated peoples: the Kurds, the Sunni and the Shiites."[29]

---

[29] Pierre Salinger with Erik Laurent, *Secret Dossier: The Hidden Agenda Behind the Gulf War.* Jonathan C. Randal's *After Such Knowledge, What Forgiveness?* is an excellent readable historical tragedy of the Kurds.

Having no homeland, it has been a national sport for the nations of the area to periodically beat the hell out of the poor Kurds. Currently, the 1991 UN Resolution bars Iraq from beating up on the Kurds, so it's Turkey's turn to slaughter them. When no one is killing the Kurds, they kill each other.

Because the Kurds are spread over five countries and are bound by traditional tribal allegiances and rivalries, it is highly doubtful they will ever gain an autonomous nation unless Iraq is carved up after the world is rid of Saddam Hussein. Nations are not prone to give away their sovereign land to 20 million people and establish neighbors who cannot agree on anything, have a history of resorting to guns to settle their internecine quarrels and change alliances to whoever is funding them or killing their enemies, including their fellow Kurds, faster than the wind changes. In short, they're like your bickering in-laws you will never get along with.

In 1996 the Patriotic Union of Kurdistan Party (PUK) had no choice but to seek support from Iran after the left leaning Kurdish Democratic Party allied itself with Saddam Hussein and invited the Iraq military to help slaughter the PUK and members of the Iraqi National Congress, run from London. All the parties were infiltrated by Iraqi intelligence; thus, it should come as no surprise that the CIA wrote off the $100 million it spent on them since the Gulf War to its "Squandered Account" and the White House claimed it only financed the group and the United States was not responsible for their actions or rescuing them, including over one hundred Kurds executed by the Iraqi forces who were on the CIA's payroll.

KURDISH ADAGE: The Kurds have no friends but the mountains.[30]

---

[30] For a brief history of the Kurds and Mid East nations, I recommend *The Times Guide to the Middle East,* edited by Peter Slugett and Marion Farouk-Slugett.

The Red Line Agreement...or the Door Closes Again

Gulbenkian didn't trust the oil companies, and was credited with the profound observation: "Oil relationships are greasy." He insisted none of the partners in the Iraq Oil Company be permitted to obtain a concession in the old Ottoman Empire he had contributed to the enterprise except through the Iraq Petroleum Company. This secret agreement admitting Exxon and Mobil barred new entrants in the Mid East and required all the partners to share in any new found oil.

Later, Gulbenkian told the story that his partners weren't sure what nations were included in the deal (they didn't), so he drew a red line on a map of the Mid East around everything but Persia and Kuwait. This pleased the Brits. APOC had the sole concession in Persia and Kuwait was a British protectorate which had agreed to grant concessions only to British companies.

Thus, the door was closed again by what was to be known as the *Red Line Agreement...but no one would know about it for several decades.*

Gulbenkian made a side deal to sell his share of the oil to CFP, a neophyte in the oil business yet to build refineries or tankers. With cash pouring in from the French, which was to earn him in excess of $20 million a year for most of his life, he bought the famous Ritz Hotel in Paris and moved into the best suite so his checks would clear faster, then he took a cruise in the Mediterranean. While sailing on the blue waters off the coast of Africa, he asked his daughter what the funny looking long ship was with its superstructure and smokestack in the stern. Until then, one of the world's richest oilmen had never seen an oil tanker.

◆ ◆ ◆

The scare that the world was running out of oil at the end of the war and into the mid 1920s waned. By the time the Red Line Agreement was signed on July 31, 1928, there was a surplus of oil coming on stream from the United States, Mexico, Venezuela, Russia and Rumania. However, the partners in the Iraq Petroleum Company controlled the price of oil they purchased in Iraq. Through a slick accounting method, called "downstream profits," they kept crude oil prices low so they could capture larger profits in their refining, transportation and marketing sub-

Red Line Agreement
July 1, 1928

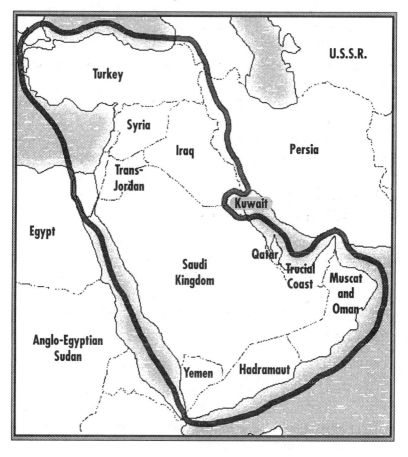

sidiaries. The members also kept production at a low level because they didn't need Iraqi oil. This made all the members happy except Gulbenkian. At the time, the Iraqis didn't know enough about the oil business to get rankled. However, the Persians became irritated when they noticed production drop and APOC had yet to let them see the books.

It was only a matter of time before the members of the Iraq Petroleum Company would have a falling out. A cutthroat price war developed between Shell and Mobil in India and spread around the world. APOC and Shell began cutting prices *and profits* in Africa and Asia. Shell became incensed after the Communists seized its oil properties in Russia and began flooding the market with cheap oil.

Shell's Sir Henri Deterding came to the conclusion that someone had to put a stop to all this dastardly competition.

Before continuing the story of "Big Oil's" shenanigans, it is worth taking a glance at these international oilmen and oil companies...

Thumbnail Sketches of Oilmen
You Didn't Learn About In School

Calouste Gulbenkian — "Mr. Five Percent" — was born in Turkey, the son of a wealthy Armenian oil merchant and banker. Educated in London in the late 1800s, he studied mining engineering and wrote his thesis on petroleum. As a young man, he prepared a study of the oil potential of Mesopotamia for Sultan Abdul Hamid. Although he never visited the area, he must have written a hell of a report or conned the old tyrant — the Sultan gave him an oil concession for the entire Ottoman Empire.

Notwithstanding his education and tremendous wealth derived from oil, he decried being called an oilman, and was quoted: "Oilmen are like cats; you can never tell from the sound of them whether they are fighting or making love."

Gulbenkian died at the age of eighty-five, attributing his longevity and sexual prowess to trading-in his mistresses when they reached eighteen.

◆ ◆ ◆

Henri Deterding was born in Amsterdam, the son of a sea captain. He was a founder of Royal Dutch and the autocratic driving force behind the Royal Dutch/Shell Group. It can also be said that he was the founder of price-fixing in the international oil industry. (John D. Rockefeller and his Standard Oil Trust deserve the credit for starting oil industry price-fixing in the United States.)

He was a close friend and business associate of Gulbenkian until they both got the hots for Lydia Pavlova, the ex-wife of a tsarist general. Deterding won her heart and took Lydia as his second wife; and Gulbenkian went back to teenage girls.

Deterding was knighted for his contribution to the British Empire during World War I. But Churchill's suspicions that he couldn't be trusted proved correct. At the age of seventy, Sir Henri was forced to retire from Shell because of Nazi sympathies. He divorced his Russian wife, Lydia, married his German secretary and moved to Germany, where he became a bosom buddy of Adolph Hitler. He died in Germany in 1939.

◆ ◆ ◆

Marcus Samuel was born in London, the son of a Jewish seashell merchant who sold trinkets and boxes decorated with shells. He developed a worldwide oil marketing organization with a tanker fleet that stunned Rockefeller's Standard Oil, which was still shipping oil around the world on freighters in blue barrels. When Shell's tankers arrived in far off places, Samuel put his oil in red barrels because there was a 10¢ deposit on them.

He was liked and respected in England and eventually elected the Lord Mayor of London. In honor of his father, the shell merchant, Samuel named all Shell's tankers after seashells. He even named one the *Clam.* Shell's marketing logo remains a scallop.

◆ ◆ ◆

Your teachers probably never taught you anything about *The Seven Sisters* either, the big oil companies that controlled 90% of the worlds exports *and all the oil in the Mid East for four decades.*[31]

---

[31] *The Seven Sisters* by Anthony Sampson is an excellent history of the international oil giants.

The Seven Sisters and the Stepsister

Enrico Mattei, the president of Italy's state oil company, Azienda Generale Italiana Petroli (AGIP), fought to obtain a share of Mid East oil for Italy during the 1950s; however, he was constantly foiled by the collusion of the major international oil companies. Frustrated, he derogatorily labeled them *le Sette Sorrelle* — *The Seven Sisters.*

As they have changed their elongated corporate names many times, they will be referred to as you know them at the pump or on your monthly credit card bill to allow the reader brand name identification along with his or her "brand loyalty" or their "brand treachery."

The purely British sister was Winston Churchill's 51% government-owned APOC, which evolved into **British Petroleum** or **BP.** Later, the British government raised its stake in BP to 56% and bought out Standard Oil of Ohio, the original core of Rockefeller's Standard Oil Trust. In 1987 Prime Minister Maggie Thatcher went on a privatization kick and sold the government's shares of which Kuwait bought 10%. **Shell** is the 60% Dutch and 40% British marketing name of the Royal Dutch/Shell Group, having purchased the 30% of Shell Oil USA it did not own.

The other five sisters *were* as American as apple pie except for one who went to bed with Saudi Arabia. Three are the offspring of Rockefeller's Standard Oil Trust, which the United States Supreme Court ordered busted up in 1911: Standard Oil Company of New Jersey, now **Exxon,** Standard Oil Company of New York, styled **Mobil,** and Standard Oil Company of California, under the **Chevron** label. The other two were the Pittsburgh Mellon family's **Gulf** and The Texas Company, trading as **Texaco,** which had the audacity to move to New York after eastern bankers stole it from its Texas founders. Texaco married Saudi Arabia's Aramco in a joint venture — Star Enterprises — and gave it half interest in three American refineries and its eastern marketing outlets as a dowry for cash and a guarantee of 600,000 barrels a day of Saudi oil.

Alert English majors may have wondered why I used the past tense *were*. There are no longer Seven Sisters. Conservative, mismanaged Gulf was gobbled up by Chevron in a "friendly" takeover rather than fall prey to the independent Texas oilman and greenmailer, T. Boone Pickens. Mattei must have been a student of Greek mythology. Zeus placed the Seven Sisters — the daughters of Atlas — in the heavens as stars, which came to be known as The Pleiades. However, only six stars are now clearly visible to the naked eye.

The Stepsister

Mattei was not entirely honest when he castigated the "Anglo-Saxon" oil companies and omitted the French Compagnie Francaise des Petroles, or **CFP**. The Italian was attempting to raise the ire of the Europeans against the British and the American sisters. Even Mattei had to admit that CFP's participation in the Red Line Agreement to keep other oil companies out of the Mid East at least made CFP a stepsister.

By the start of World War II, the Seven Sisters and French stepsister controlled all the oil in the Mid East through consortiums.

Figure 1

SEVEN SISTERS' MID EAST CONCESSIONS — 1950

Country	Concession	Ownership	Percent
Iran	BP	British Govt.	56
		Burmah Oil Co.	22
		Privately held	22
Iraq	Iraq Petroleum Co.	BP	23 3/4
		Shell	23 3/4
		CFP	23 3/4
		Exxon	11 7/8
		Mobil	11 7/8
		Gulbenkian	5
Saudi Arabia	Aramco	Chevron	30
		Texaco	30
		Exxon	30
		Mobil	10
Kuwait	Kuwait Oil Co.	BP	50
		Gulf	50
Bahrain	Bahrain Pet. Co.	Chevron	50
		Texaco	50
Qatar	Petroleum Dev. Ltd.	Same as Iraq	
U.A.E.*	Various	Same as Iraq	

*Seven separate concessions.

Achnacarry Castle — "Let's Make a Deal"

Shortly after signing the Red Line Agreement, Sir Henri Deterding invited a few chums to Achnacarry Castle in the Scottish Highlands for a "bit of grouse hunting and fishing." Included in the guest list were Boss Teagle of Exxon, Sir John Cadman of BP, William Mellon of Gulf and Colonel Robert Stewart of Standard Oil Of Indiana (Amoco.)[32] Although William Mellon didn't have a title like Boss, Sir or Colonel, his uncle was Andrew Mellon, the Secretary of the Treasury, which gave him standing — the Mellons were personally wealthier than the other hunters and fishermen.

Teagle reported: "The hunting was lousy." But that's not the real reason they spent a week in Scotland. On rainy days (it rains a lot in Scotland), they drafted what was to become known as the *Achnacarry* or *As Is Agreement*. The purpose of this 17 page agreement was to stamp out competition. Although the agreement would have been a violation of United States antitrust laws, the Webb-Pomerane Act of 1918 permitted U.S. companies to fix prices and market shares outside the United States. Winston Churchill, now the Chancellor of the Exchequer — the Brits' elegant title for the equivalent of the United States Secretary of the Treasury — supported the price fixing because the British government owned controlling interest in BP.

The As Is Agreement did precisely what it was named. It eliminated competition by setting quotas on the members' respective worldwide market shares as they were in 1928. The market shares were to remain as is. Price competition was to be avoided. Any member producing in excess of its quota was required to sell it to a fellow member.

The members agreed to assist each other by allowing the use of their facilities to other members on a favorable basis and to draw or exchange crude oil or refined products from the nearest producing field or refinery to the market, regardless who owned it, and split the savings on transportation costs.

---

[32] Amoco never became a big player in the Mid East, deciding to concentrate its oil exploration and production in Venezuela. Colonel Stewart lost his job when the Rockefellers, who still owned a big chunk of stock in Amoco, discovered Stewart had swindled Amoco out of a couple of million bucks in a fraudulent oil deal.

United States antitrust laws were avoided by a slick maneuver. World crude oil prices were to be based on the price of oil on the Texas Gulf Coast *plus* the cost of transportation to the market *regardless* of where the oil was produced. The so-called *Gulf plus* pricing system added substantial margins — United States prices were the highest in the world. Further, the distance between the Texas Gulf and Europe was greater than between the Mid East and Europe. Coupled with their agreement to swap oil production for delivery to the nearest market, the Gulf plus price added to their hidden profits. In oil industry jargon, the profit split was termed a "transportation differential." Behind closed doors it was laughingly called *phantom freight.*

For obvious reasons, an agreement to limit production and observe quotas in various countries couldn't be called the *As Is Agreement.* Instead, it was given the clubby name "Pool Association." Although there were suspicions, no one knew about the agreement until it was uncovered by the Federal Trade Commission in 1948 — twenty years later — and its contents weren't revealed to the public until 1952. As you can imagine, when the news broke, there were a lot of pissed off Arabs and consuming nations.

To make the deal work, the hunters and fishermen in Scotland had to entice other major companies to join in the scheme. Seventeen American oil companies went along with the deal, which included not selling crude oil or finished products to outsiders. The big outsider that had to be induced was Russia, which joined the conspiracy in 1929 and obtained a guaranteed share of the British market. The deal didn't have to be explained to the Russians, who were experts on planned economies and prices fixed in Moscow. The French set company import quotas in France so its neophyte CFP stood a chance competing with the big boys.

One of the Sisters not joining the club in the beginning was Chevron, sitting in San Francisco. Chevron believed it was too far away to fool with Arab oil and it had no European market. It also knew that one of its Sisters would cheat if the price was right. It was inevitable a few members would renege on the Achnacarry Agreement. Russia, as expected, adhered to the agreement only when it suited them. Regardless, the cheap Mid East oil was a bargain. It provided the original members with a reserve they could turn on and off as needed and added fat profits under the Gulf Plus pricing system.

But there were stumbling blocks facing the Achnacarry Agreement members. The Russians were a minor obstacle compared to the American independent oilmen, especially the wild and wooly Texans.

Dad Joiner -- The Texas Wildcatter Who Found Too Much Oil

On October 3, 1930, Columbus Marion "Dad" Joiner, a crippled, seventy-year old con man, brought in a gusher in East Texas that was to revolutionize the oil industry. It was the largest oil discovery in the history of the United States until the Prudoe Bay, Alaska, strike in 1967.

At the onset of the Depression, tens of thousands of job-hungry Americans flocked to the East Texas oil boom. By the end of 1931, the East Texas Oil Field — "The Black Giant" — was capable of producing one-third of America's insatiable thirst for oil.

Dad Joiner searched for three years where no one believed there was oil. The major oil companies laughed at the old man who financed his one rusty, rickety drilling rig with watered stock by romancing widows whose names he picked out of their husband's obituaries. Many said that Dad didn't believe it himself when they discovered the old swindler had sold 350% of his wildcat venture.

Before Exxon's Texas subsidiary, the Humble Oil Company, could say "Armadillo," the oil field was found to stretch forty-five miles north to south and fifteen miles wide and small independent oilmen had enveloped the area to the chagrin of "Big Oil," which controlled oil prices. The flood of oil out of East Texas sent the price in Texas plummeting from $1.15 to 10¢ a barrel in less than a year. Big Oil couldn't maintain the *Gulf Plus* prices they had fashioned in far off Scotland, and a battle erupted between the Texas independent oilman and Big Oil. *Big Oil* was easy to define: it was big and not from Texas. It included Texaco, The Texas Company that moved its headquarters up north to the damnyankee (one word) big city of New York to be run by the bankers and Wall Street manipulators who had caused the Depression.

As oil prices dropped, the independents answered by drilling more wells and producing more oil. At the low point, spot crude oil prices were reported at 2¢ a barrel. However, the oil boom drove the price of hamburgers from 5¢ to 25¢ in East Texas to the consternation of the oilmen and smug delight of economists who believed in the law of supply and demand.

When it became apparent that the oil was being pumped out too fast, allowing salt water aquifers to intrude in the pool of oil and causing the pressure that brings the oil to the surface to drop, Big Oil, supported by large independent oilmen, came up with a system to conserve the precious resource. The strategy was called *market demand prorationing*. In brief, market demand prorationing allowed the Texas Railroad Commission to determine how much oil could be absorbed in the market as well as maintain the proper pressure in the field. The scheme was to raise the price of oil back to $1 a barrel.

The Texas Railroad Commission concluded that each well should be limited to 225 barrels a day, later reduced to 75 barrels a day. Gushers capable of producing 20,000 barrels a day were restricted to a trickle of their capacity. This suited Big Oil, that by now had bought out many bankrupt independent oilmen and had access to cheap oil in Venezuela and the Mid East. Small independents, with only a few wells, continued to go under.

The independents fought back. Being a rugged and sneaky lot, they took to bootlegging oil, known as "hot oil" to distinguish it from bootleg booze during our nation's ill-fated attempt at Prohibition. Hot oil was trucked and shipped in illegal pipelines from oil wells hidden in the woods, barns and phony houses to the major cities in the west and midwest. At its peak, it was estimated that bootleg oil hit 500,000 barrels a day — one-sixth of America's demand.

With the oil fields in chaos, two evil characters straight out of an early "B" western emerged. The Governor of Texas, Ross Sterling, declared martial law and sent the National Guard to shut down the East Texas oil field. The fact that the governor was the former president of Exxon's subsidiary, Humble Oil, made more than a few Texans give him the evil eye. Needless to say, he wasn't re-elected — defeated by a woman, Miriam A. "Ma" Ferguson.

Leading the National Guard troops was the short, mean-spirited General Jacob Wolters who, when he wasn't quelling riots during the

Depression, was general counsel of Texaco. Big Oil had stacked the deck. And, so he wouldn't feel lonely, Wolters brought as his adjutant a Gulf Oil Company officer. It is a known fact that short men in uniform try to bully people, termed a Napoleon complex after another short general. His authority over the oil field included barring the hookers from parading down the street advertising the wares in colored beach pajamas. (His explanation was that the nearest beach was over 100 miles away.) General Wolters believed he was above the law and refused to obey a Federal court injunction holding the Texas marketing demand prorationing law illegal.

Meanwhile, political mischief abounded in the nation's capital. Big Oil, represented by the American Petroleum Institute (API), recommended that the total oil production of the United States be limited to the amount produced in 1928, the year of the Achnacarry Agreement quotas, which the government and independent oilmen knew nothing about. Several years earlier, the API had opposed any restrictions on oil production, even for conservation purposes. Boss Teagle thought he could get rid of the pesky independents by convincing the Department of the Interior that the Federal government should buy the East Texas oil field and hold it as a reserve in case of another war; however, the Federal government was broke and couldn't afford it.

In 1932 the United States levied a 21¢ a barrel tariff on imported oil to shut out foreign oil. This devastated Venezuela, which had been exporting over half its oil to the United States, but helped raise the price of oil in America.

It wasn't until 1934, when Congress passed the Connelly Hot Oil Act, did bootleg hot oil dribble to a stop. The Act's sponsor was Senator Tom Connelly of Texas, a strong supporter of "Dollar Oil." The Act prohibited the interstate transportation of hot oil, defined as oil produced contrary to the conservation laws of a state, and brought in J. Edgar Hoover's FBI to enforce the law. In the interim, oil wells and pipelines in East Texas had been dynamited and a few folks shot at.

National market demand prorationing was enforced under the Interstate Oil Compact Act of 1935, which allowed the oil producing states to designate production quotas based on a monthly estimate of the nation's demand by the Bureau of Mines — the same faceless bureaucrats who said America was running out of oil a few years earlier. In later years, when the omniscient Washington number crunchers erred on the low side, needless spot shortages occurred and prices rose, even though the nation had an oil surplus.

In the end, Big Oil got its way and increased oil prices during the Depression with the help of President Roosevelt's National Recovery Act, later declared unconstitutional by the Supreme Court. The Supremes also ruled that Governor Sterling's declaration of martial law was a subterfuge and unconstitutional.

As if by magic, the price of oil jumped to over $1 a barrel and made a passel of Texas millionaires and added to the profits of Big Oil.

Market demand prorationing also had an insidious result: It forced a period of artificially high oil prices in the United States that would last forty years...until the Arab embargo during the 1973 Yom Kippur War. In 1973 the United States government would think of other ways to inflate the price of oil.

QUIZ: What happened to Dad Joiner?

Answer: It has been told that Dad Joiner died penniless because, at the age of seventy-three, he married his twenty-three year old secretary and gave her a charge account at Neiman Marcus. The old wildcatter lived to eighty-seven, still looking for another gusher, but finding only dry holes. In the meantime, the Bible and poetry quoting romancer set a record at the Dallas Public Library for taking out the most books.

Dad never had to face his irate investors in court for selling 350% of his company and one oil lease eleven times. He was "rescued" by a womanizing bigamist and professional gambler who won his first oil well in a game of five card stud a few years earlier. Rumors were that the gambler, H.L. Hunt, flimflammed Dad out of his oil field and eventually paid him less than $2 million. H.L. Hunt parlayed Dad's bonanza into one of America's greatest fortunes and died an old right-wing curmudgeon, leaving his 14 legitimate and not-so-legitimate children to squabble over a couple of billion dollars.

QUIZ: What does the Texas Railroad Commission have to do with oil?

Answer: The Texas Railroad Commission regulates everything important in Texas, including railroads, corporations and oil. It was the brainchild of Governor Jim Hogg in the late 1800s during his successful attempt to recover the railroads the damnyankee carpetbaggers stole after the War Against Northern Aggression. The TRC did such a bodacious job, it was given authority to straighten out all sorts of commercial rambunctiousness.

In 1988 when oil prices hit rock bottom, a member of the TRC showed his independence by going to Vienna, Austria, to chat with OPEC and determine if Texas should join the OPEC cartel.

Governor Hogg became wealthy as a result of oil investments, which shouldn't come as a surprise. He also had a peculiar sense of humor. He named his daughter *Ima Hogg*.

9

ARABIA — We're Awash in Oil, But We Still Want More

The period of gloom of the 1930s was known as the Great Depression. It could just as well have been called *BLASH — Buy Low And Sell High.* A rudimentary principle of the petroleum industry is that crude oil is a wasting asset. Even the largest oil fields will become depleted sometime in the future and have to be replaced. By obtaining oil reserves during a depression at low prices, they can only multiply in value during the next inevitable boom. And once you have them, never, never, never let them go. It is also elementary that you don't let another company get more oil reserves, especially if you want to stay a member of the Seven Sisters...*But, as we all know, sisters fight on occasion.*

Bahrain & Abu al-Naft — "The Father of Oil"

I'm the Sheik of Araby
Your love belongs to me
At night when you're asleep
Into your tent I'll creep...

Abu al-Naft — "The Father of Oil" — to the people of the Arabian peninsula, was a New Zealander, Frank Holmes. Although his homeland was a British dominion, the rough-hewn mining engineer was an outsider who didn't wear a British old school tie. What earned him the wrath of the chaps in London was that Holmes planned to sneak across their Red Line and, to make matters worse, bring Americans.

Holmes had heard tales of oil seeps in the tiny archipelago of Bahrain off the coast of Saudi Arabia. To his consternation, Sheik al-Khalifa thought oil would mess up his tents and carpets and wasn't interested.[33] The sheik told Holmes, if he really wanted to do something helpful, he should drill for water. To get on the sheik's good side, Holmes drilled for water and found it. That pleased the sheik so much, he gave Holmes an oil concession.

Before Holmes and the sheik finished celebrating, the British told them that as a protectorate the sheik wasn't supposed to even discuss important things like oil unless they approved. Holmes thought the problem was solved when APOC scoffed at the silly thought that there was oil in Bahrain. Unable to sell his concession in Britain, Holmes took the next boat to America, the land of opportunity. Boss Teagle of Exxon turned down Holmes' offer to sell the concession for $50,000 because Bahrain was included in the Red Line Agreement and he couldn't find the 240 square mile sheikdom on a map. Teagle later called his failure to grab the concession a "billion dollar error."

Gulf Oil purchased Holmes' concession because it couldn't depend on its sisters to sell it oil at a decent price. However, as Gulf was purchasing crude from the Iraq Petroleum Company, it was subject to the Achnacarry and Red Line Agreements and prohibited from obtaining a concession without the members' participation. Although the Red Line clique didn't believe there was oil in Bahrain, they didn't want to be proved wrong and allow Gulf to flood the market. Rather than gamble, Gulf turned chicken and sold the concession to Chevron.

The first thing the British told Chevron was that it wasn't a British company, which Chevron already knew because it was from California. What the British meant was that in all the treaties with its protectorates, there was fine print that required all oil companies to be British. After

---

[33] A Sheik from the al-Khalifa clan has ruled Bahrain since 1782 when the Persians pulled out. Shah Muhammad Pahlavi tried to grab it back in 1971 after the British withdrew from the Persian Gulf, but the United Nations said: "Hell, no!" President Nixon told the Shah he couldn't manage Iran and already had enough oil.

Gulf War veterans liked Bahrain. The people are friendly and they serve booze. The Saudis go there when they need a drink because of the ban on alcohol in their country.

a year of screaming "Open door" by the United States government, the British caved in. To save face in London, Chevron was compelled to organize a subsidiary in Canada, which was still British enough to have a picture of King George V on its money. In addition, Chevron was required to funnel all its correspondence and contracts through the British government's agent in Bahrain. After all, ignorant little Bahrain still needed British protection.

It took six years for Frank Holmes to run the British gauntlet and the obstacles of the Red Line Agreement, but it was worth it. Chevron discovered oil in Bahrain in 1932.

Bahrain is too small to become a major world oil producer or join OPEC. Its importance in Mid East oil history is that it was the first place the Red Line Agreement was broken.

BEDOUIN PROVERB
As any Bedouin will tell you:
Once a camel's nose slips under the tent,
his big ass is sure to follow.

Kuwait...Abu al-Naft Slips in Again

Gulf also acquired Frank Holmes' concession in Kuwait, which was outside the Red Line Agreement.[34] Earlier, APOC had drilled a dry hole 30 miles south of Kuwait City and claimed the oil rights to the entire protectorate. Although APOC believed there was no oil in the tiny sheikdom, BP's predecessor didn't want any Americans mucking about in its backyard. However, Sheik Ahmad al-Sabah was adamant. Kuwait's pearl industry was going down the tubes because Japanese cultured pearls had displaced natural pearls on the world market and he saw his neighbors in Bahrain reaping revenues.

As expected, APOC pulled the old British-nationality-clause-in-the-protectorate-treaty routine in an attempt to keep the Americans out. Gulf sent in its big gun, Andrew Mellon, the former Secretary of the Treasury and now United States Ambassador to Great Britain. The pressure of the multimillionaire politician who bankrolled Gulf worked. This time, APOC was not going to let the Americans run off with the family jewels. In December 1934 the Kuwait Oil Company was organized with APOC and Gulf each holding a 50% interest. The British insisted on administering the agreement, but the Americans did the drilling. In February 1938 oil was discovered near where the British had drilled a dry hole.[35]

---

[34] Sheik Mubarak the Great, an ancestor of the al-Sabah family who still run Kuwait, took control of the sheikdom by murdering his two brothers in 1896, then asserted his independence from the Ottoman Empire in 1897 by asking for British protection. Thus, Kuwait, which means "Little Fort," was a protectorate and not part of the Ottoman Empire when Gulbenkian obtained the oil concession.

[35] To this day, few international oil companies consider drilling for oil without a few Texans or Okies to work their drill rigs or American technology. If any Englishman denies this, send them to the old Duke's Wood oil field in England. There stands a statue to 40 Okies who brought in an oil well during World War II after the Brits failed. (Today, American roughnecks are in Russia trying to straighten out the mess they made of their oil fields.)

Saudi Arabia...The Jewel of Arabia

Like his neighbors, King Ibn Saud was broke. Income from the pilgrimages to Mecca and Medina fell drastically during the Depression. One day, one of his advisors advised (that's why Ibn Saud had advisors) that he should grant an oil concession. There were tales of oil seeping from the ground in the Al Hasa area. Like Bahrain's Sheik al-Khalifa, Ibn Saud was more interested in water. This may explain why his advisor brought in Charles R. Crane, an American plumbing manufacturer and former member of the King-Crane Commission, for additional advice.

Another explanation why the advisor invited the American was that he despised British Mid East policy. The advisor's Islamic name was Abdullah; however, he had been christened Harry St. John Bridger Philby in England. Philby had been a member of the British Indian Civil Service during the carve up of the Middle East and believed the Arabs were getting screwed. Not only did he think it was tawdry for the Indian Civil Service to run the Middle East from some 2,000 miles away, he believed it asinine to appoint Faisel, a Hashimite, king over a mob of unruly Mesopotamians and the creation of Transjordan didn't pass the sneer test. This is important, because Mid Eastern potentates, and Ibn Saud was no exception, insist their advisors hold the same opinions they do. Philby had favored an independent Iraqi republic and later resigned as the British Resident in Transjordan over the mess his countrymen had made of things. This turned out to be a good career move. Philby soon found there was more *baksheesh* being an advisor to a king than as a civil servant. In no time he became the local Ford dealer and obtained the kingdom's wireless (radio) concession.

Philby and the American plumber convinced Ibn Saud his interests would be best served by Americans. After the Bahrain oil strike, Chevron contacted Philby with a proposal for an oil concession. Philby was a masterful "double advisor." First, he advised the Iraq Petroleum Company of Chevron's interest in order to jack up the price in a bidding war. Then he became Chevron's secret paid advisor during the negotiations, while still holding the ear of Ibn Saud as an advisor.

The Iraq Petroleum Company never had a chance. They came to the bargaining table with only £10,000 in British sterling (US$50,000–

roughly equivalent to US$2 million in 1998). Chevron offered £35,000 in gold, another £20,000 in 18 months and £100,000 if they found oil. Philby had let Chevron in on the fact that Ibn Saud liked gold and thought paper money with a picture of a king on it wasn't worth a hell of a lot, especially if it wasn't a picture of King Ibn Saud.

In March 1938 an immense oil field was discovered in Saudi Arabia. Chevron's oil concession would later be found to cover one-fourth of the world's oil reserves.

◆ ◆ ◆

In the United States, things had hit rock-bottom. Roosevelt was busy worrying about the Depression and trying to pack the Supreme Court that was ruling everything he was doing unconstitutional. Roosevelt neglected the Mid East, which was being inundated with Germans, Italians and Japanese hot to obtain oil to fuel the next war. It wasn't until 1939 that the United States accredited its ambassador in Egypt to handle Saudi Arabia in his spare time. American diplomacy in the Mid East would be left to the oil companies to fret over for several years.

Chevron protected its interests by exercising the option to expand its concession to cover the entire nation of Saudi Arabia (over three times the size of Texas) and shutting out the Axis powers and British. To establish world markets for its Saudi oil, Chevron invited its sister, Texaco, to share its interests in Saudi Arabia and Bahrain, partly in exchange for Texaco's international markets. So there would be no confusion with the other Sisters' Red Line Agreement, they formed a Blue Line Agreement to divvy up their world markets and combine their holdings "East of Suez" under a new company called Caltex.

The United Arab Emirates...The Forgotten Jewel
The United Arab Emirates is a federation of seven emirates that were British protectorates until 1971 when the British pulled their troops out of the Persian Gulf. The UAE is another small part of the world overlooked by chroniclers of petroleum history and high school teachers. The cause stems from the fact that nothing interesting happened there and they can't pronounce the names of the emirates or the myriad

of sheiks. High school teachers are also concerned how to pronounce the name of former Abu Dhabi Sheik Shakhbut's name without upsetting classroom decorum.

TRIVIA QUIZ: Name the seven UAE emirates:

Answer: Abu Dhabi*, Dubai*, Sharjah,* Ajam, Fujairah, Ras al-Khaimah* and Umm al-Qaiwain. [*The emirates with oil.]

One of few places the Red Line Agreement worked was the UAE, beginning with Abu Dhabi, the largest of the emirates and the one with the most oil. With the aid of the British government, when the Iraq Petroleum Company noticed American oil companies grabbing chunks of the Arabian peninsula, it swooped in and signed each of the emirates to 75 year concessions. As each emirate is a sovereign state, individual concessions had to be signed. The fact that the borders were not settled added to the confusion. There are still border disputes among the emirates and with Saudi Arabia and Iran, the latter involving islands in the Gulf.

As the Iraq Petroleum Company did not need more oil, it sat on the concessions from 1935 until after World War II. Oil was not discovered until the 1950s; however, the wells were enormous. After the Red Line Agreement was set aside, several American independent oil companies obtained concessions in the late 1950s.

Sheik Shakhbut of Abu Dhabi was the Jed Clampett of the Mid East — an Arab hillbilly millionaire. When flying to London, he carried his lunch in a sack. (In his defense, Arab sheiks had been known to poison their relatives in order to take over a sheikdom.) It wasn't until he couldn't figure a way to stop the mice nibbling at the paper money stored under his bed did he put it in a bank. Afterwards, he regularly woke the manager of the local bank in the middle of the night to ensure that his money was still in the vault.

The UAE, about the size of Maine, contains 10% of the world's oil reserves, ranking it third behind Saudi Arabia and Iraq. The United States has a defense pact with the UAE for reasons not necessary to explain. It's about time high school teachers begin mentioning the place, even if it's only to say that the people are very rich and friendly or that their Ford may have gasoline refined from UAE crude oil in its tank.

And Last, So They Won't Feel Left Out, Little Qatar

Slightly smaller than Connecticut, barren Qatar sticks into the Persian Gulf like a fat thumb. Britain believed it should have been a dependency of Bahrain and would never amount to anything when it was signing up protectorates in the 1800s, so it waited until World War I before allowing Qatar to become a protectorate. The Ottomans occupied Qatar until Ibn Saud took over the adjacent Al Hasa region of Saudi Arabia in 1914 and things got too hot for them. Even into the 1920s, Qatar paid Ibn Saud 100,000 rupees a year not to overrun it, which doesn't say much for British protection. Qatar didn't settle its border dispute with Saudi Arabia until 1992. The sheikdom has been run by the al-Thani clan since the nineteenth century.

The Iraq Petroleum Company claimed Qatar was covered under the Red Line Agreement and entered into an oil concession in 1935 before the Americans could sneak under the tent flap. As in the case of neighboring UAE, the members of the Red Line Agreement didn't need additional oil at the time, and Qatar's oil fields were not developed until after World War II.

The tiny sheikdom, with a population of about 500,000, generally follows its big brother, Saudi Arabia's, lead and has never been an influence on OPEC policy. It was the first nation to join the five OPEC founding members. Actually, it wanted to be a founding member, but the original OPEC members thought it was too small and would make OPEC look "too Arab."

No one knows much about Qatar. In 1994 it signed a defense pact with the United States; therefore, it's highly possible that American troops will visit there someday and tell us about the place when they get back.

QUIZ: Should the UAE and Bahrain have been covered under the Red Line Agreement?

Answer. Both were British protectorates when Gulbenkian obtained the concession from the Sultan of Turkey and were not part of the Ottoman Empire. Perhaps the real question should be: Did Gulbenkian pull a fast one on the Seven Sisters?

QUIZ: What happened to Philby?

Answer: Philby died wealthy after surviving a midlife primitive circumcision when he converted to Islam. The operation was a success. He fathered a child at sixty-five by a slave girl Ibn Saud gave him. Philby was a crotchety dilettante. The British jailed him in London for anti-British statements during World War II after being tipped off by Ibn Saud who, while not crazy about the British, believed advisors shouldn't go around shooting their mouths off. After Ibn Saud died, his son, King Saud, spit on Philby and expelled him for criticizing his new brand of government.

Philby was given the name Abdullah by Ibn Saud, which means "slave of God." Most Saudis thought his name should have been Abdulqirsh (slave of sixpence).

Philby's double-dealing and enmity towards his native England was passed on to his infamous son, Kim Philby, the British double-agent who defected to Russia. Kim's treachery makes the CIA's recent blunders look like petty larceny. Why the Brits allowed Kim Philby into their secret service (MI6 as any reader of Ian Fleming knows) is a mystery. He was known to have pro-communist leanings while a student at Cambridge, old school tie notwithstanding.

QUIZ: Who was Charles R. Crane?

Answer: Crane inherited the toilet and sink business but never bothered to run it. This gave him time to travel around the world and meddle in international affairs. He was qualified to serve on the King-Crane Commission in the Mid East by reason of being the largest contributor to Wilson's 1912 campaign for President.

In 1909 President Taft withdrew his appointment as American minister to China because of his open hatred of "Japs and Jews." He turned down Wilson's offer of ambassador to Russia because "the Bolsheviks shouldn't be taken seriously." Crane met Hitler and struck up a friendship because of their common hatred of the British, French and Jews. It should come as no surprise that the King-Crane Commission report opposed a Jewish homeland in Palestine. **Short answer: Crane was a meddlesome, bigoted nut.**

10

WORLD WAR II — You Can't Trust Anyone

War came to the Middle East in October 1940 when Italian Air Force planes took off to bomb the British protectorate of Bahrain. The Italians missed and bombed the Dhahran oil field in neutral Saudi Arabia. Unable to defend the entire Gulf area or rely on Italian navigation, the British plugged the wells in Kuwait and the Americans shut in production in Saudi Arabia except for small quantities sent to the Caltex refinery in Bahrain.

The British had their hands full. They had to recruit troops of their vast empire from faraway India, Canada, Australia and New Zealand to protect the British Middle East at the beginning of the war, Churchill called "our darkest hour."[36] Britain's toughest adversary, Germany's Desert Fox — General Irwin Rommel, battled Britain back and forth across the North African Desert until he ran out of oil and was defeated at El Alamein in October 1942. The British took Tripoli in January 1943, and Italy surrendered in September.

---

[36] In England, during the dark hours when they were getting the hell beat out of them, they sang a song to raise their spirits, *There'll always be an England*. The author's mother told him that the next line should be: *as long as Scotland's there.* Her family, the MacDonalds, swear what makes Great Britain is Scotland. As career Scottish soldiers, they fought in India, Afghanistan, South Africa, France, Egypt, Iraq and Sudan (also Yorktown, but she seldom mentioned that). Soon, the Scot's bagpipes would be heard at El Alamein and Tripoli.

The French were knocked out of the war in June 1940 after the fall of Paris, not that they would have been much help. In the end, the French were their usual double-crossing selves by fighting against the Americans in North Africa and adding to the confusion...But there's a few more things you must learn if you are to understand why the British and French compounded the Mid East's distrust.

At times, the chaos in the Mid East was tantamount to anarchy. The history and characters are so little-known, the names and events have yet to appear on the TV quiz show *Jeopardy*. The obscure war in the Mid East, few Americans knew or cared about, added to the Mid Easterners' paranoia towards the European empires and was fought to maintain French and British colonies and *oil*.[37] Iran's output of 215,000 barrels a day of oil exceeded Nazi Europe's entire production, and Iraq's wells turned out another 75,000 barrels a day.

General Rommel never reached Egypt's oil fields yielding 20,000 barrels a day. If he had, King Farouk of Egypt said he would welcome the German Afrika Korps, "as liberators from the unbearable brutal British yoke." In February 1942 the British surrounded Farouk's palace with tanks in order to convince the King to appoint a pro-British prime minister. Farouk had made the mistake of believing the British meant it when they granted Egypt independence in 1922. Crowds cheering Rommel's name in the streets of Cairo told the British Tommies they were not welcome in their former protectorate. The pro-German feeling of the Egyptians was not deep-rooted, the people were anti-British. Throughout the Middle East, there were not many friendly faces.

---

[37] The critical role played in World War II is documented in the highly readable and fascinating *Oil & War* by Robert Goralski and Russell W. Freeburg.

ANWAR SADAT

During the war Egyptian Army Captain Anwar Sadat was arrested and interned* after being caught working for the pro-German underground. He would become Egypt's President between 1970 and 1981, and continue to give the British fits. He even gave the United States and Russia a hard time. Later, he made friends with Menachem Begin of Israel and Jimmy Carter and signed a peace treaty with Israel. Americans thought he was a great guy, but his Arab neighbors believed he committed treason against all Arabs. For this, he would die from assassins' bullets. [*Interned is a polite way of saying he was a prisoner of war.]

Syria & Lebanon — The French Connection

After the fall of France, the German puppet government in Vichy granted the Germans the rights to Syrian and Lebanese airfields. British Anzac and Indian and Free French forces moved quickly to take Beirut and Damascus, control the airfields and secure the oil pipeline from Iraq to the Mediterranean at Haifa. General Charles de Gaulle's representative signed an armistice with the Vichy forces on July 14, 1941 — Bastille Day — and gave the Vichy troops in Lebanon the option to be part of Free France or Vichy. The celebration of France's national holiday was climaxed by 32,000 of the 38,000 troops (84%) *siding with Vichy.*

In November 1941 Lebanon proclaimed its independence. The Syrians overwhelming voted for self-government in 1943. However, the will of the Arab people didn't count any more then than it did after World War I when the Americans took a poll indicating French were not welcome. It was the gall of de Gaulle that mattered. De Gaulle didn't like the way the elections turned out and continued to command the *Troupes Spéciales.* This was not something on a French menu, but the local army of "Special Troops" controlling the area. But, as they weren't French, he didn't trust them. When Germany surrendered in May 1945, de Gaulle sent French colonial troops and bombarded Damascus to let the Syrians know he was boss. Nothing had changed after either war. The French soldiers were back home drinking Burgundy and Bordeaux with their *cheries*. De Gaulle ordered the same black Senegalese troops France sent after World War I to do France's dirty work.

De Gaulle irritated the Allies, who were tired of the war. The United States and Russia had already recognized Syria and Lebanon as independent nations and did not want to look fickle in the eyes of the Arabs. The British were fed up with France not doing anything to win the war and claiming the spoils. President Truman told de Gaulle to get his troops the hell out of the Mid East or he would cut off French aid and sensitive parts of their anatomy. De Gaulle caved. Thus ended the French connection.

QUIZ: The term "Vichy" means:
(a)   A sparkling mineral water, like Perrier, drunk by snobs in America because it sounds French.
(b)   A soup made from potatoes, leeks and cream served in fancy French restaurants.
(c)   The German puppet capital of unoccupied France during World War II.

Answer: Both (a) and (c). Do not complain to the *maitre d'* that your *vichyssoise* is cold.

QUIZ: President Truman's opinion of de Gaulle and France was:
(a)   "I don't like the son of a bitch."
(b)   "De Gaulle is a psychopath."
(c)   "The French ought to be taken out and castrated."

Answer: All of the above are quotes by "Give 'em hell Harry."

Iraq — The Land of British-made Kings

Iraq gained its independence in 1932, but remained under Britain's watchful eye, having promised "full and frank consultations with Great Britain in all matters of foreign policy."

Faisel Hussein died in 1934 and was succeeded by his son, Ghazi I, who was killed in a sports car accident. Rumors abounded that the British knocked him off, but the truth is Ghazi was a terrible driver. Next in line was four year-old King Faisel II, who took over with his uncle as regent and a pro-British premier who had served with Faisel I and Lawrence of Arabia during the Arab revolt. The premier was aptly named Nuri al-Said because whatever Nuri al-Said said, little Faisel did.

With the German Luftwaffe next door in Damascus, pro-German Iraqis seized power under Rashid Ali al-Gailani, who took over as prime minister in 1941. The American ambassador wrapped Nuri al-Said in a rug, tossed him in the back of his car and drove him to safety. Rashid Ali refused to permit the British to land forces to protect the oil fields and ordered arms and supplies from the Axis powers. Hitler's offer of help had strings attached Rashid Ali couldn't swallow and still keep his head — the Führer insisted on the rights to Iraq's oil in perpetuity.

As only the chaps at Whitehall can do with aplomb, the British government announced: "The prime minister of Iraq no longer enjoyed its confidence." The Brits determined they didn't need permission to protect *their* oil fields. Within a few weeks, British Indian troops and the Arab Legion from Transjordan, led by General J.B. Glubb, made short work of Rashid Ali's forces.[38] Rashid Ali fled to Saudi Arabia and the protection of Ibn Saud who refused the British request to extradite him, letting King George VI know that Ibn Saud was king in the desert. Later, when Rashid Ali made too many pro-German remarks, Ibn Saud booted his tail to Germany.

---

[38] English General Glubb headed Transjordan's Arab Legion, the only Arab mob in the Mid East that looked like an army because it was financed and trained by the British. The Bedouin Arab League forces called the general *Glubb Pasha*. Loosely translated, Pasha means "Big Shot."

With the oil fields secure and Nuri al-Said back telling the boy king what to do, Iraq declared war on Germany in January 1943. It can be safely said (a phrase the author picked up from debating his English friends), the Iraqis were more anti-British than pro-German. This is understandable, Americans didn't like being run by King George III in 1776. Britain lost more friends in Iraq during the war, and young king Faisel II would have to learn to walk a narrow path or he could lose his ~~ass~~ head.

ARAB DOUBLE-DEALING

This is not a reprint of a previous sidebar. As he did in 1918 when he approached the Turks for a better deal than the British offered, Nuri al-Said contacted the Germans in 1941 offering an alliance between Iraq and Germany, but the Germans didn't believe anything Nuri al-Said said.

Iran — the Shah's Ill Health

Germany's advance towards the Russian oil fields in the Caucasus in 1941 made Iran's position along the Russian border strategically important to the defense of Russia and Iranian oil. Iran declared its neutrality and refused the joint British and Russian request to ship arms to Russia over the Trans-Russian Railway. Iran's appeal to the United States to protect its neutrality fell on deaf ears. Roosevelt had already agreed to provide Lend-Lease supplies to Britain and Russia, and wondered what 3,200 German "advisors" were doing in a neutral country.

A little country like Iran should have known better than to deny the British Lion's and Russian Bear's offer of protection. The Allies invaded from the north and south as they did in World War I, promising to "defend Iran by all means at their command from aggression." In a hurry to protect AIOC's refinery at Abadan from unfriendly forces, the British sank several Iranian naval ships on the way. Britain's Indian troops also shot several British AIOC oil field workers because they "looked German."

Three weeks later Shah Reza Pahlavi abdicated because of failing health brought on by pro-German ingestion and was exiled to Mauritius in the Indian Ocean. The Shah's twenty-three year-old son, Muhammad Reza Pahlavi, was installed on the Peacock Throne in his place. This was necessary because the Allies needed a new shah to declare war on Germany in 1943 and a safe place to hold the first meeting between Roosevelt, Churchill and Stalin. The young Shah knew he was in good hands when the Big Three promised to defend the "independence, sovereignty and territorial integrity of Iran."

By the end of the war 30,000 American troops were stationed in Iran to handle the war supplies and food going to Russia over the "Bridge of Victory." As for the Iranians, food shortages amounting to a famine devastated parts of the land. Britain didn't win many friends in Iran, who was supposed to be their ally. Roosevelt sat mute...after all, the British were in charge of the Mid East.

After the war, young Muhammad had his father's body exhumed in South Africa, where he had died in exile, and returned to Tehran for a grandiose funeral. Not to be outdone by earlier great Persian kings, such as Darius and Cyrus the Great, equestrian statues throughout Iran proclaimed his father, "Shah Reza Pahlavi the Great."

Saudi Arabia & Other Oil Hot Spots

At the outbreak of World War II, the myopic United States State Department had a total of *thirteen* people, including four secretaries, in the Near East Division to handle an area that stretched across North Africa and the Middle East to the border of India. The staff could only boast of three who could speak Arabic. The American ambassador to Egypt, who served as a part-time ambassador to Saudi Arabia, had little knowledge of Ibn Saud. If he had, he may have learned that Ibn Saud had made a deal with Germany for 4,000 rifles. Ibn Saud was adept at being all things to all men.

After promises of riches when Chevron struck oil in 1938, Saudi oil production was drastically curtailed during the war and few Muslims were making the pilgrimage to Mecca. Ibn Saud was in dire need of cash. In July 1941 President Roosevelt turned down a request for aid to Saudi Arabia with a curt handwritten note to his Lend-Lease administrator:

> "Will you tell the British I hope they can take care of the King of Saudi Arabia. This is a little far afield for us."

The British responded by pouring £6 million (US $30 million) into Ibn Saud's strongbox during the next two years. Naturally, while chatting over tea with the king, the British mentioned that it would be keen if they could have a piece of the Saudi Arabian oil concession. Chevron and Texaco had no choice but to advance US$4 million during both 1941 and 1942 against future oil royalties. The oil companies, America's only real ambassadors in the Mid East at the time, explained to Ibn Saud that the British aid was merely a method of funneling United States aid to the king, which made sense — America was granting aid to Britain.

Ibn Saud spent the war building palaces for himself and his dozen or so sons who had reached marriageable age. As one can imagine, a king who fathered 43 sons and scads of daughters and had palaces full of wives had lots of personal expenses.[39]

— Back in America...

No one listened to the oil companies' pleas for help to protect their Mid East interests until Secretary of the Interior Harold Ickes, America's Petroleum Co-ordinator for National Defense, calculated that America was running out of oil. Senator Owen Brewster of Maine was alarmed and thundered: "Before another generation comes of age, America will become a mendicant for petroleum at the council tables of the world." Roosevelt, a Harvard man, was one of the few Democrats who knew

---

[39] Historians differ as to the number of daughters sired by Ibn Saud. Baby girls weren't as important as boys and no one kept close track of them. Maybe Ibn Saud didn't want to admit he was producing daughters over sons at a rate of five-to-one. Ibn Saud had 23 wives, but never more than four at once.

what Republican Senator Brewster meant by mendicant. Immediately, FDR started dumping Lend-Lease into Saudi Arabia with both hands to the tune of $33 million during the next two years, as he didn't want to be outdone by the British.

Harold Ickes wasn't known as the "Old Curmudgeon" because he was a sweet old man.[40] The Democrat Roosevelt selected Ickes for the job because he was a tight-fisted Republican lawyer. Ickes, a caustic, power-hungry New Dealer, thought the United States government needed Saudi oil more than the oil companies. After the Lend-Lease money started flowing into Saudi Arabia, Ickes told Chevron and Texaco of his grand plan for the government to buy them out and name the new company the American-Arabian Oil Company.[41]

Everyone opposed Ickes' idea — Republicans, Democrats, Communists and the oil companies, especially Exxon and Mobil who were planning on joining their sisters in the world's greatest oil bonanza. As a result, the United States government didn't go into competition with British Petroleum; however, the Brits were suspicious.

— Between the Allies, Things Were Getting Sticky...

Churchill viewed American political strength being built up in the Mid East as a play for British oil and wrote Roosevelt that the British were "being hustled." Roosevelt wrote back that the British were trying to "horn in" on America's oil concessions in Saudi Arabia. This was plain talk for two erudite masters of the English language. Being astute politicians and not wanting to quibble over the issue in the middle of a war, Churchill and Roosevelt decided to split up the Mid East oil.

In August 1944 the United States and Great Britain signed the Anglo-American Petroleum Agreement setting out their control of Mid

---

[40] The bombastic Ickes' autobiography was accurately entitled *Autobiography of a Curmudgeon.*

[41] The oil companies knew that the name would never fly with Ibn Saud. When they reorganized the company, it was named *Arabian-American Oil Company — Aramco.*

East oil and establishing the International Petroleum Commission whose task it would be to estimate worldwide demand and allocate production quotas. In effect, it was the legalization of the Achnacarry Agreement, with the twist that eight member governments of the Commission would determine the oil demand and allocation.

The American oil industry didn't trust Roosevelt's "New Dealers," Churchill's "Limeys" and Stalin's "Commies" to allocate *their* oil and attacked the Anglo-American Petroleum Agreement with both feet when it was submitted to the Senate for confirmation. Eyebrows were also raised when rumors began flying that the government was planning to file an antitrust action against the Seven Sisters for doing what the agreement provided — fixing oil prices through the allocation of world-wide oil production. Roosevelt withdrew the Agreement in disgust when the Democrat controlled Senate said it reeked of colonialism and hanky-panky. Of interest, no one asked the Arabs if America and Britain could split up *Arab oil.*

WHO NEEDS ARAB OIL?

In 1944 I.F. Stone, writing for *The Nation*, correctly predicted, "If we are to depend on Arabian oil, we must be prepared to defend the sea and air routes over which it must travel to the United States." But, he was dead wrong in claiming estimates of Saudi oil reserves were exaggerated at between 3 and 20 billion barrels. In 1996 they were estimated at 258 billion. In 1995 the Saudis produced 3 billion barrels.

— Meanwhile, in Venezuela, Things Were Hopping...

Ickes' 1943 "strategic shortage" in the United States encouraged the burning of foreign oil. Venezuela was a secure source that could be transported safely to America now that German U-boats were no longer engaged in a turkey shoot sinking American and British ships within sight of Atlantic City and Miami, whose local governments refused to dim their lights and made the ships sitting ducks.[42]

America's neighbor to the south, Mexico, was not regarded as stable because of its nationalization of American and British oil companies in 1938; and its national oil company, Petróleos Mexicanos (Pemex), was renown for its corrupt and inefficient management and a bloated work force to keep down political unrest. It still is. Also, the fact that Mexico sold oil to Germany and Japan at the beginning of the war was regarded as a violation of Roosevelt's "Good Neighbor" policy. It was America's first lesson that good neighbors meant little when balanced against oil. Only American warships convinced Mexico to become a good neighbor again.

Shell and Exxon, whose oil properties had been confiscated in Mexico, were the major oil producers in Venezuela and astute when it came to negotiating oil concessions with the crooked dictator, Juan Vicente Gómez. They had learned that dictators didn't always stay bought and new regimes had a bent for changing the rules. Shell and Amoco built refineries off the coast of Venezuela on Aruba and Curaáao in the Netherlands Antilles, where they were safe under Dutch control. Everyone was happy when Gomez died. Venezuela's oil appeared secure...for awhile.

The Venezuelans recognized the United States' need for its oil and decided it was a good time to renegotiate the under-the-table deals made with the dead dictator. Roosevelt wanted the oil so badly he sent Herbert Hoover, Jr., the son of the Republican President he whipped in 1932, to act as a consultant to the Venezuelans.[43]

---

[42] German U-boats sank 2,275 Allied merchant ships along the Atlantic seaboard. Michael Gannon's best seller, Operation Drumbeat, is a fascinating story of Germany's U-boats and the toll they took on shipping. Only the building of an oil pipeline from the Texas oil fields to refineries in New Jersey, constructed in less than one year, guaranteed American's Northeast a secure supply of oil.

[43] Herbert Junior was a geologist. Herbert Senior was the first and probably last mining engineer to be elected President. Both were bright and boring. Herbert Senior and his wife spent their evenings translating the first known treatise on mining from Latin to English: De Re Metallica, a thriller by Georgius Agricola written in 1556.

This was a major turning point in the world of oil company national concessions. Venezuela negotiated a *fifty-fifty split* of the companies' profits by increasing its royalties and taxes. To insure there would be no reneging and chicanery after the war, the legislature enacted a law confirming the fifty-fifty split.

Juan Pablo Pérez Alfonzo of the minority Acciòn Democràtica Partido (liberal-socialist), who understood petroleum accounting, calculated the split closer to sixty-forty in favor of the oil companies, but no one listened. After the Acción Democrática coup of 1945, Pérez Alfonzo became the Minister of Development and demanded a real fifty-fifty split based on the companies' refining and marketing profits, then demanded another first from the concession holders — that Venezuela market its share of oil on the world market through a Venezuelan state oil company.

In 1948 there was another coup, and Pérez Alfonzo and his party were run out of the country. The new regime liked the fifty-fifty split, but wasn't about to try to run an international oil company. Pérez Alfonzo would have to wait until the next coup to return from exile for the third time to raise hell with the oil companies. He wasn't missed by the American oilmen. Independent oilman, William F. Buckley, was glad to see the Acción Democrática Partido get chucked out on its ear, as they were nothing but "anti-American communists." [44]

— In Far Off Indonesia, Things Were Going To Hell...

The Japanese occupied what was then the Dutch East Indies from early 1942 until Japan surrendered in 1945. President Sukarno declared independence from the Dutch and gave the nation the name Indonesia, which eventually included Bali, Java, Borneo, Sumatra and many other beautiful islands. What Sukarno didn't count on was the British troops

---

[44] Buckley sent his son, William F. Buckley, Jr., to Yale where he learned a lot of big words. Junior became a conservative columnist who also believes that communists, socialists and liberals are still up to no good.

on hand to accept Japan's surrender didn't think it was cricket for the Indonesians to declare independence before the Dutch troops arrived. The British held the nationalists, some of whom called themselves "Oil Freedom Fighters," at bay until the Dutch troops took over. It would not be until 1949 that the Indonesians and Dutch formed a short-lived "union." In 1956 Indonesia pulled out of the union and became an independent nation.

Historians may differ as to why the British refused to recognize the Indonesian declaration of independence from the Dutch in 1945. The author has two theories. Prior to the war, the Dutch East Indies produced 170,000 barrels of oil a day, more than all of Europe outside of Russia. Shell controlled the majority of the production, which was split sixty-forty between Dutch and British interests. They were not about to turn over the oil to a bunch of uncivilized islanders still rumored to eat people. It may also be explained by the British red faces, who figured they owed the Dutch an explanation for the surprise capture of the Sumatran oil fields by Japanese paratroopers in February 1942. The Japanese had sneaked up on the Dutch in 50 British Royal Air Force planes, the Brits were careless enough to permit the Japanese to capture in Malaysia, before the Dutch could defend themselves and sabotage the oil fields. The Indonesian oil fields were Japan's greatest source of oil during the war.

Indonesia is another place history and geography teachers skip. Most Americans haven't a clue about the place unless they subscribe to *National Geographic*. It is an enchanting country full of charming, friendly people, but don't drink the water.

— By 1945, the Good Guys Knew They Were Going to Win...

Roosevelt, Churchill and Stalin met in Yalta to divide up Germany and hand over Poland and a few other small nations to Russia. As they were leaving, Roosevelt casually mentioned to Churchill that he was going to drop by and see Ibn Saud on his way home.

The meeting was secret. No one knew about it in Saudi Arabia except Ibn Saud and several of his sons holding top positions in government, as he was to meet Roosevelt in the Great Bitter Lake in the Suez

Canal within range of German bombers.[45] As the King's caravan read-
ied to depart Jeddah for Mecca on a family outing, Ibn Saud told the drivers to head for the dock. One hour later, Ibn Saud's entourage boarded the American destroyer *USS Murphy.*

America's new Ambassador William A. Eddy, who acted as
Roosevelt's translator, was respected and liked by the Saudis, and negotiated the last minute details.[46] The first snag was limiting Ibn Saud's party
of over 100 to 12, as demanded by the destroyer's cramped quarters. The
king compromised and boarded an entourage of 48. Actually, the number
mattered little. Ibn Saud lived in a tent on the deck, foregoing the offer of
the captain's cabin while his servants slept under the sky. A flock of sheep
being herded towards the gangplank as a hospitable gift to feed Ibn Saud's
new American friends was corralled after Eddy explained that it was
against regulations for sailors to eat lamb that wasn't kept in ice boxes. Ibn
Saud thought Americans were crazy to eat meat that wasn't killed the same
day and insisted on bringing seven sheep for his family. The Saudi cooks
managed to find room in the gun turrets and munitions storage bulkheads
to tend their fires for roasting lamb and brewing coffee during the two day
journey without blowing up the ship. During the evenings, the captain and
Eddy entertained Ibn Saud on deck. If Ibn Saud knew that his sons were
below deck with the American sailors watching films of girls running
around in scanty bathing suits, he never mentioned it.

◆ ◆ ◆

---

[45] Ibn Saud couldn't tell all his sons or it would not have been a
secret. He had thirty-four living at the time, ranging in age from two
to forty-three, and little boys can be blabbermouths. The al-Saud
family still doesn't trust outsiders. The country is run by his sons and
grandsons. They stopped counting the princes of Ibn Saud's lineage
when they reached 5,000.

[46] Colonel Eddy, a former OSS officer, was fluent in Arabic and
gained the admiration of the young Saudi princes by translating the
rules of basketball into Arabic. He was a true "Arabist" and resigned
from the State Department in protest of Truman's pro-Jewish policy
in Palestine. He was buried in Lebanon in 1962 with "U.S. Marine"
engraved on his headstone.

Ibn Saud took an immediate liking to Roosevelt, who gave him a DC-3 aircraft in return for jeweled daggers and rings presented by the King. When the old men compared their infirmities — Roosevelt's confinement to a wheelchair due to polio and Ibn Saud's limp from bullet wounds — the President gave the King his spare wheelchair, a gift Ibn Saud never forgot. He kept the wheelchair on display in his quarters for the remainder of his life. Nor did Ibn Saud forget Roosevelt's forbearance from smoking and drinking alcohol in his presence. (Ibn Saud caught FDR sneaking a cigarette behind a dinghy.)

The two leaders talked about American oil interests, but little of interest was recorded. What is remembered is Ibn Saud's answer to FDR's question what to do with the homeless Jews of Europe after their terrible suffering and persecution during the holocaust. The wise King advised: "Give them and their descendants the choicest lands and homes of the Germans who oppressed them." Roosevelt, the consummate politician, agreed to consult with Ibn Saud and other Arab leaders before doing anything on the "Palestine question."

— Three Days later, Churchill Visited Ibn Saud...

Churchill was nonplused. Roosevelt had out-foxed him. His £500 box of perfumes was no match for Roosevelt's gift of a DC-3, so glib Winston promised Ibn Saud the first Rolls Royce off England's assembly line after the war.

The Prime Minister could not live without whiskey and pungent cigars, and told the King as much as he imbibed and puffed in his presence. His big stick approach on the Jewish homeland nettled Ibn Saud. Churchill's bragging what he had done for the Arab kings — the Husseins — who Ibn Saud defeated twenty years earlier as enemies and religious usurpers went over like a camel turd in the royal tent.

Churchill paid little heed to Ibn Saud's deep religious belief that Allah had condemned the Jews for their persecution of 'Isa (Jesus) and their rejection of the Prophet Muhammad. As far as giving Arab land to the Jews, Ibn Saud believed: "It is easier to give away other peoples' countries...and not so dangerous," and wondered aloud: "How would the people of Scotland react if the English gave away Scotland?"

When the custom made Rolls Royce finally arrived, Ibn Saud was insulted again. The regal throne-like back seat brought a sneer — only women sat in the back. Men sat in the front next to the driver on the *right, never the left*. The British steering installed on the right was too much. Without entering the $75,000 Rolls Royce, Ibn Saud gave it to his brother who liked to drive.

— In America, the Land of the Free and the Home of the Brave...

State Department diplomats celebrated. The Americans had won the diplomatic game against the British in their Mid East backyard. Ibn Saud was firmly pro-American and anti-British. The King had returned to Saudi Arabia victorious after receiving a pledge that there would be consultations on a Jewish homeland in Palestine...

Two months later, Roosevelt died.

There was not even a charade of consultations. The new President, Harry Truman, said: "I have to answer to hundreds of thousands who are anxious for the success of Zionism; I do not have hundreds of thousands of Arabs among my constituents [voters]."

Truman was aware the Republican contender for the Presidency, Governor Tom Dewey of New York, had come out for a Jewish state in Palestine. Almost all of Truman's advisors were opposed to a Jewish state, including such notables as Secretary of Defense George Marshall, Secretary of State Dean Acheson, George Kennan, Charles Bohlen, James Forrestal, and Loy Henderson. It was not anti-semitism, they knew the strategic value of Arab oil and they were embarking on a worldwide struggle against the Soviet Union. The only top advisor favoring a Jewish state was Truman's political advisor, Clark Clifford.

As an absolute monarch, Ibn Saud knew that promises of a ruler are buried with him. It also meant that Ibn Saud was not bound by any promises or allegiances.

QUIZ: Did Churchill make a serious mistake in praising the Hashemite kings — the Husseins?

Answer: He must have guzzled too much brandy in the hot sun. The Husseins had battled the al-Sauds for over a century. When the British installed Sherif as King of The Hijaz and his sons, Abdullah and Faisel, in Transjordan and Iraq, Ibn Saud blamed the British for surrounding Saudi Arabia with his enemies.

The straw that broke the proverbial camel's back occurred when Sherif proclaimed himself Successor to the Prophet, Guardian of Islam — the leader of the Muslim world. Not only was Ibn Saud pissed off, Muslims in Indonesia said it was an effrontery to Allah and Indian Muslims claimed it was another dirty British trick to take over the Islamic world. It was Ibn Saud's devout *Ikhwan* warriors who captured the Holy Cities of Mecca and Medina to rid the Arabian peninsula of the Husseins.

Another reason Churchill may have goofed is that he forgot the Husseins like to call themselves Hashemites.

11

PEACE IS HELL — The Mid East Heats Up in the Cold War

"Sherman was wrong...*peace is hell,*" President Truman told the annual Gridiron Club dinner in December 1945. In the aftermath of World War II, the United States faced the challenge of converting its wartime economy and working for an elusive peace. America could no longer sit isolated concerned only with domestic problems. It soon found itself involved in a Cold War with the Soviet Union and the establishment of a Jewish state in Palestine.

For the first time, America was forced to become entangled in the mysterious Mid East, and it would have to play the cards it was dealt. The hatred of the French in Syria and anti-British sentiment in Egypt, Iraq, Iran and Palestine could easily be transferred to the United States and allow the Soviets to fill the power void in the region.

It was also the first time the Arabs flexed their muscles. In March 1945 the League of Arab States was organized by the seven Arab states that had gained their autonomy — Egypt, Iraq, Saudi Arabia, Transjordan, Yemen, Lebanon and Syria. Today, the Arab League has twenty-two members. including the semi-autonomous state of Palestine. No study of the Mid East is complete without mentioning the Islamic revival enhanced by the creation of Israel by the Christian West that had subjugated the Mid East. Islam was and remains the most potent seed of Arab unity and nationalism. The vast majority of the populations of nine of the eleven OPEC nations is Islamic. Only Venezuela has a large Christian majority. Nigeria is divided 50% Islamic and 40% Christian. Wags whispered that former OPEC Christian members, Ecuador and

Gabon, felt uncomfortable hearing the call of the *muezzin* five times a day. It was the Organization of Arab Petroleum Exporting Countries (OAPEC), *not OPEC,* that embargoed oil shipments to the United States for its support of Israel during the Yom Kippur War in 1973,[47] while Soviet Russia was supporting the Arab nations.

In 1948 the United States became a net importer of oil. The post-war production of automobiles brought a 50% increase in the demand for gasoline and almost doubled the price of crude oil between 1945 and 1948. America's seesaw foreign policy in the Mid East would be governed by the Cold War with Soviet Union, Israel and oil, a strategy of reaction rather than action, lacking honesty and foresight.

Iran — The First Cold War Crisis

Before the war wound down, Shell and Mobil approached Shah Muhammad Pahlavi for oil concessions in the northern provinces of Iran not covered by the British AIOC concession. Hot on their heels came the Soviet Union demanding the same deal. The Russians had a leg up. They had armed the Iranian Tudeh party (communists) in the north and the Tudehs in the Majlis were demanding a political balance between Russia and the West. However, the Shah was able to postpone any decision until the end of the war because of the presence of British and American troops.

As required by the Big Three's agreement, the United States and Britain withdrew their forces from Iran after the war, but the Soviet army hung around continuing to arm the Tudeh and pressing the Azerbaijan and Kurdistan provinces under their control to secede from Iran and join the Soviet Union. Stalin responded to Truman's demand that the Soviets remove their troops with the first big lie of the Cold War. He explained that the Baku oil fields lay perilously close to the border and he was concerned about hostile action by Iran against the Soviet Union, which was like saying the Green Bay Packers were worried about their next game with the St. Mary's High School Junior Varsity.

---

[47] The author is guilty of showing bias by referring to the 1973 war as the Yom Kippur War. To the Arabs, it is the Ramadan War — named after the Islamic holy days.

In early 1946 Soviet troops marched towards the Iranian city of Tabriz. Iran, with the support of the United States, brought the matter to the United Nations. This was the first issue brought before the fledgling UN and marked the first Cold War confrontation. Russia evacuated its troops, in part, because Iran's Prime Minister, Ahmad Qavam, promised the Soviets he would present the Majlis with an agreement for a joint Iranian-Soviet company (51-49% in favor of the Soviets) to exploit the oil in the northern provinces. However, after the Soviets left, Qavam double-crossed them and pressed the Majlis to reject the agreement.

Unfortunately for Qavam, Shah Pahlavi thought he was getting too big for his britches. With the help of his twin sister, Princess Ashraf, who many thought far brighter than the Shah, he convinced the Majlis to remove Qavam, then appointed a prime minister he could control. In February 1949 the Shah was shot and wounded by an alleged communist, which gave the Shah the opportunity to proclaim martial law, outlaw the communist Tudeh party and exile his enemies. On a brief wave of public support, he dissolved the Majlis and became as pompous and autocratic as his father. The British preferred dealing with one leader rather than a myriad of ministers and the Majlis because it was more expedient and cut down on the number of bribes. Truman, while favoring democracies, believed a stable Iran could only be accomplished by the authoritarian Shah... Boy, was that a mistake!

America's first announced step into the region was the Truman Doctrine, extending aid to Greece and Turkey to stem the Soviet Union's influence. Britain, without the financial resources to protect its former domain and oil after the costly war, was replaced as the Mid East's sugar daddy by the United States, which poured aid into Iran under the Truman Doctrine.

QUIZ: Name the three biggest lies told by dictators, kings, generals and shahs.

Answer:
(1) As soon as I restore law and order, we'll have free elections and reestablish democracy.
(2) It's untrue that I've jailed and killed my political opponents.
(3) I need American planes and weapons to protect my country from Soviet communist enslavement **and/or**
(3) I need Soviet planes and weapons to protect my country from American capitalistic oppression.

The Palestine Partition — You Can Slice a Pie But Not a Stew

A great deal has been written about the Palestine partition and the creation of Israel — all biased on one side or the other. Like everyone else, historians have prejudices. A detailed history is beyond the scope of this short text. I urge the reader to search out *all* viewpoints before making up his or her mind. The following is but a brief recap of the events surrounding the partition.

The world was aware of the potential Arab-Israeli conflict as the war drew to an end. In March 1945 the threat was made clear when Saudi Arabia, Egypt, Syria and Lebanon signed a pledge to take military action to protect Arab interests in Palestine and Transjordan. The State Department and the Joint Chiefs of Staff warned, if the United States antagonized the Arabs, the Soviet Union would evolve as the major power in the Mid East, with dire consequences for the control of Arab oil.

Everyone, especially the Arabs, underestimated the will and strength of the Jews. The Haganah, the illegal Jewish army which later became the Israeli army, was supported by two terrorist groups, the Irgun Zvai Leumi and Stern Gang, who waged guerilla war and terrorist attacks against the Arabs and the British forces occupying the Palestinian mandate. The terrorism included the massacres and destruction of Arab villages, the assassination of the British minister in Cairo and the blowing up of the British headquarters in the King David Hotel in Jerusalem. That's not to say the Arabs were not resorting to terrorism, just that the Jews were much better terrorists in those days.

QUIZ: What is the definition of a terrorist?

Answer: It depends on whose side you're on and who history declares a terrorist. The Stern Gang, which assassinated Count Folke Bernadotte, the United Nation's Mediator for Palestine, was led by Yitzhak Shamir; and the Irgun was headed by Menachem Begin, who led the massacre at Deir Yassin where over 250 Arab Palestinian civilians were slaughtered. Both were later elected prime minister of Israel. Yassir Arafat's Palestinian Liberation Organization were initially considered terrorists. So was Anwar Sadat, the anti-British captain interned by the British, who later became President of Egypt. *Later, all four received the Nobel Peace Prize.* This is logical. Alfred B. Nobel, the Swedish philanthropist who established the Nobel Peace Prize, earned his vast fortune from his invention of dynamite.

In November 1947 the United Nations voted to partition Palestine into separate Arab and Jewish states by a vote of 33 to 13 with 10 abstentions. Jerusalem was to remain under an international mandate. The disorganized, voiceless Palestinians said it was morally wrong for the UN to give away their homes and land. Israel's David Ben-Gurion publicly accepted the resolution while planning to overrun the borders. Abdullah Hussein of Transjordan, with the knowledge of the British, spoke out against the partition, but conspired with Israel to allow him to snatch the West Bank. The United States and Russia supported the partition scheduled to take place when the British pulled out on May 14, 1948. Under political pressure during an election year, Truman recognized the new State of Israel eleven minutes after it declared its independence.

U.N. Partition of Palestine
November 1947

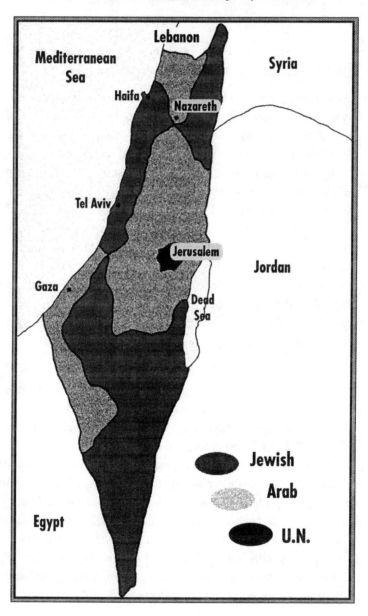

The partition was doomed to failure from the start. It was a gerry-mandering of three separate enclaves for both the Arabs and the Jews that barely connected at two points, making it ripe for confrontation. Both Jews and Arabs were to be cut off from relatives, places of employment and where they prayed by artificial boundaries marked by fences and enemy soldiers.

The day after the British evacuated, 20,000 troops from Egypt, Syria, Iraq, Transjordan and Lebanon moved towards the new Jewish state with the "declared intention" of occupying the area of Palestine allotted to the Arabs. They were no match for the 60,000 men and women army of Israel, which didn't desire an Arab army next door and grabbed everything that had been partitioned by the UN as an Arab state except the Egyptian-controlled Gaza Strip and the West Bank that Israel permitted Abdullah to steal.

To the world, it appeared a tiny nation of 600,000 defeated 40 million Arabs surrounding it. An estimated 800,000 Palestinian Arabs fled their homeland to neighboring Arab nations. All the Arabs managed to accomplish was cut off the oil pipeline from Iraq to Haifa, Israel, and Egypt refused passage of oil tankers and merchant shipping through the Suez Canal. The Arab world, humiliated by its defeat, settled back to wait for a strong Arab leader and their next opportunity to reclaim what they believed were Arab lands.

PALESTINE BOX SCORE

The Palestinians refer to the 1948 War as *al-nakba* — the disaster. The UN partition granted Israel 54% of the land even though the Jewish community held only 8% of the land and made up one-third of the population. However, the worst was yet to come. When the war was over, Israel held 75% of the land and Transjordan had grabbed the West Bank. Also, Israel and Transjordan had split up Jerusalem, which was supposed to be under UN trusteeship. Israel annexed the lands captured during the war. There was nothing left and 800,000 Palestinians had been forced to leave the country...**The score was Palestinians ZERO!**

Notwithstanding UN Security Council Resolution 194 in December 1948 calling for a return of the lands to the Palestinians, the exiled Palestinians have never been permitted to return to their homes, nor have they been compensated for their losses. *Thus, more seeds of violence were sown, and the reason for Palestine's objection to today's Jewish settlements in Palestine becomes obvious. Today, when one looks at the maps of land being negotiated for the Palestinians, compare it to the map of the UN partition and ask what happened to the rule of law and where have the honest peace brokers and high-minded UN delegates been since 1948?*

QUIZ: World opinion and the UN would object if —
(a) The United States conquered and annexed Mexico because it wanted its oil and tacos?

(b) Iraq invaded Saudi Arabia in order to obtain Mecca and Saudi oil to pay reparations for its invasion of Kuwait?

(c) Norway invaded Britain and demanded a partition giving it Scotland's oil and golf courses?

Answer: Yes. The difference between Israel conquering Palestine and annexing the lands is there is no oil in Palestine.

Transjordan — The Mouse (Double-crossing Rat?) That Roared Too Loud

Transjordan was mocked by Palestinian Arabs and Jews, scorned by the Bedouin, and laughingly called "Churchill's Inspiration" in Britain. As you may recall, Lawrence found "Abdullah too clever." Abdullah was not content with the vacant lot called Transjordan. His father had been promised the title of "King of all the Arabs." In 1942 Abdullah and Nuri al-Said, the premier of Iraq and brains behind

Abdullah's young grandnephew, Faisel II, developed the concept of a "Greater Syria and Fertile Crescent," by unifying Palestine, Transjordan, Lebanon and Syria, with the eventual inclusion of Iraq. They also fantasized that Saudi Arabia might join, an indication they may have been smoking some of that fine hashish grown in the Bekaa Valley in Lebanon.

After Britain abolished its Transjordan mandate in 1946, Abdullah promoted himself from amir to king. The Soviets said he was a British puppet and vetoed Transjordan's admission to the United Nations. To show its independence and save face, Transjordan entered into an alliance with Iraq, signed by Abdullah's grandnephew, which didn't pass the international laugh test.

Transjordan's Arab Legion, led by British General Glubb, was the only thing that resembled an Arab army when the Arabs marched toward Palestine in 1948 and the Arab Legion occupied the West Bank, with approval of Israel. Clever Abdullah had been discreetly chatting with the Zionists about taking the West Bank and ridding it of its Arab nationalist leaders who might cause them both trouble.

In 1949 Abdullah renamed his wasteland The Hashimite Kingdom of Jordan. Next year he incorporated the West Bank into Jordan. Every Arab nation objected but Iraq, which didn't surprise anyone, and became suspicious when Abdullah began talking about the idea of a Greater Syria and Fertile Crescent. Only the Palestinians seemed to care that the UN partition established *independent Jewish and Arab states.*

In July 1951 Abdullah was assassinated in a Jerusalem mosque by a Palestinian in protest of his "Greater Syria" pipedream. Abdullah was succeeded by his son, Talal, who was anti-British and declared insane (assuming the two factors weren't related). Talal's seventeen year-old son, Hussein I, took over in 1952. Because he witnessed his grandfather's assassination and what happens to you if you step out of line, he has tried desperately to be friends with his neighbors. A kingdom between Syria, Iraq, Israel and Saudi Arabia requires its king to sit on a throne that inflicts its monarchs with perpetual hemorrhoids.

Periodically, Hussein declares that he hates whichever nation it's politically correct to despise at the moment. The following month, he will declare his love and desire for peace with the same nation. Jordan's barren desert is not economically important and has more than its fair share of Palestinian refugees, but no oil. Hussein is the longest reigning monarch in the Mid East because no one else wants the impossible job. Jordan survives on handouts from its wealthy Arab neighbors and American charity we call foreign aid. Maybe that's why he needed an American woman for his queen.

12

BIG OIL — Big Profits, Big Secret

Offshore Oil — Who Owns It?

Oil had been produced from piers jutting into the Pacific waters near Santa Barbara, California, since 1897, although California didn't get around to issuing state oil and gas leases until 1921. In 1933 Interior Secretary Harold Ickes refused to issue federal oil and gas leases in the tidewater because "title to the soil under the ocean within the three-mile limit is vested in the State of California." Texas and Louisiana had leased 5 million acres in the Gulf of Mexico by 1945.

However, all that happened before the technology was developed to drill in over 200 feet of water and Ickes told President Truman that America was running out of oil and the federal government should protect the nation's oil against state giveaways to Big Oil. Alarmed, Truman signed a proclamation in 1945 declaring that the federal government owned all natural resources in the seabed of the continental shelf, extending 200 miles under the sea. Up until then, most nations claimed a three-mile jurisdiction based on the international law concept conceived by Hugo Grotius that a nation's waters extend one marine league, the effective range of the most powerful cannons in 1625.

Politically, all hell broke loose after the Supreme Court upheld the federal government's claim to the seabed. Twice Congress passed legislation granting the states the rights to the natural resources within the three-mile limit, which Truman vetoed. Eisenhower, a Texan, sided with California, Texas and Louisiana and signed the legislation when it was enacted for the third time in 1953.

Today, the technology exists to drill in waters over one mile deep between the three-mile limit and 200 miles claimed by the federal government. California, the pioneer in drilling in state waters that bitterly contested the federal government's claim in the Supreme Court, banned drilling in its state waters in 1995 for environmental reasons. The deep waters of the United States are one of the few virgin places left to explore but, because of the great expense, it is strictly the domain of Big Oil.

What does Truman's ocean grab have to do with the Mid East? After much debate, in 1958, the Geneva Convention adopted Truman's concept of ownership of the continental shelf. The 200 mile claim still causes disputes in the Persian Gulf, which is less than 400 miles wide in most locations and dotted with tiny islands no one believed were worth claiming until oil was discovered in the waters. One quarter of the oil produced by Saudi Arabia and the UAE comes from offshore wells.

Big Oil Forms the World's Biggest Oil Conglomerate — ARAMCO

Chevron and Texaco had more oil in Saudi Arabia than they could market and their concession could be in jeopardy if they did not satisfy Ibn Saud's demands for increased production. They turned to their two sisters, Exxon and Mobil, to help market the oil and invest in a pipeline from Saudi Arabia to the Mediterranean, the Trans-Arabian Pipeline or "Tapline." Exxon and Mobil jumped at the opportunity, but an old bugaboo reared its ugly head — they were members of the Red Line Agreement.

For the first time, the Americans said that the Red Line Agreement was illegal. While Shell and AIOC said they would work out a deal, the French company, CFP, and Gulbenkian cried foul and went to court. It mattered little that Gulbenkian's concession was limited to the former Ottoman Empire and Saudi Arabia disputed it was part of the empire, slippery Gulbenkian had drawn a red line around Saudi Arabia. The French and Gulbenkian settled the law suit after Ibn Saud said he didn't care what the courts said, he would only permit American oil companies in Saudi Arabia. At least, American oil company diplomacy was successful in the Mid East.

The Exxon and Mobil purchase of 40% of the new company, Arabian-American Oil Company (Aramco), cost $470 million in 1947 dollars, dwarfing today's multi-billion dollar mergers and buyouts accomplished with smoke, mirrors and junk bonds. At the last minute, Mobil chickened out and only took 10%, leaving the other sisters with 30% each. It was a decision Mobil would regret. Exxon gobbled up half of Mobil's share and laughed on the way to the bank for many years.

◆ ◆ ◆

The monumental Aramco deal insured America's mastery in international oil for decades. What the United States forgot was that it would not have been possible except for Ibn Saud...and worse...that Ibn Saud was America's best friend in the complex Arab Mid East.[48]

The Barren Desert — Where the Seven Sisters' Profits Are Discovered

In 1948 Saudi Arabia and Kuwait decided to grant oil concessions in the Neutral Zone, carved out of the desert by the British in 1922 when they couldn't agree on borders. The U.S. State Department insisted only American independent oil companies should bid on the oil rights, fearing they might be criticized for favoring the American Seven Sisters, who had already slurped up the choice oil spots in the Mid East. The Alice in Wonderland State Department should have realized that American independent oilmen were like Mad Hatters — berserkers craving to drill for oil anywhere, anytime at any cost.

Kuwait granted its concession to a consortium of large independent oil companies, Phillips, Sinclair and Ashland, called Aminoil — short for American Independent Oil Company. The price shook up the Sisters. The price was $7.5 million in cash, 15% of the profits and a royalty of 35¢ a barrel with an annual minimum guaranteed payment of $625,000. In addition, Aminoil threw in a million dollar yacht for the Amir of Kuwait, Ahmad al-Sabah — for Kuwait's *half* interest in 2,000 square miles of desert.

---

[48] No study of the Mid East and Saudi Arabia is complete without reading *The Kingdom: Arabia & the House of Sa'ud* by Robert Lacy.

The price for Saudi Arabia's *half* interest in the Neutral Zone went for a higher price. J. Paul Getty's Getty Oil paid $9.5 million up front, a 55¢ a barrel royalty with a minimum guarantee of $1 million a year, plus promises to construct schools and furnish the Saudi Army with gasoline. Obviously, Ibn Saud wasn't aware Getty had authorized his negotiator to go as high as $10.5 million.

The Seven Sisters went bonkers. The crazy American independents paid too much! Aramco's royalty payable to Saudi Arabia was 33¢ a barrel; the Iraq Petroleum Company and AIOC were paying roughly 16 1/2¢ a barrel to Iraq and Iran; and Gulf and AIOC were low at 15¢ a barrel to Kuwait. To the independents, it was worth it. They hit a bonanza. If the Neutral Zone — the size of Delaware — were a nation, it would rank nineteenth in world oil reserves. The great disparity in royalty payments did not go unnoticed by the Mid East nations.

J. Paul Getty went on to become America's first billionaire and the stingiest billionaire of all time after installing a pay phone for his guests in his English manor house.[49]

FAMOUS AND NOT-SO-FAMOUS GETTYISMS:

"The meek shall inherit the earth, but not the mineral rights."

"My wives married me; I didn't marry them." (Getty should know, he had five wives.)

"If you can count your money, you don't have a billion dollars."

"When I'm thinking about oil, I'm not thinking about girls."

---

[49] For a best seller about the outrageous life of J. Paul Getty, I recommend Robert Lenzer's *The Great Getty.*

In Venezuela, They Knew About the Oil Company Profits

The Venezuelans also knew about the United States' Internal Revenue Code. In their dealings with American oil companies to obtain a fifty-fifty split in 1943, they discovered that taxes paid to Venezuela by the oil firms were credited against the taxes due the United States, while royalties were merely deductions in the computation of net income. The "foreign tax credit" for taxes paid Venezuela allowed Venezuela *and* the oil companies to reap greater net profits from the pockets the American taxpayer.

The State Department justified the oil tax credit windfall on the ground it avoided the problem of having to go to the Congress to request foreign aid for the poor nations — political doublespeak. Those Mid East nations not spending their oil revenues building huge military forces, such as Saudi Arabia and Kuwait, were soon floating in an ocean of petrodollars, a term coined by economists for the obscene profits from their sales of crude oil.

Venezuela sent a delegation to the Mid East to explain the fifty-fifty concept and the United States tax laws in 1949. The Venezuelans were not merely being nice guys, they were concerned that cheap Mid East oil was replacing their oil in the United States and Europe, and they needed to maintain their oil exports. They were also aware that Mid Eastern crude oil was cheaper to produce and a higher quality than the heavier Venezuelan crude oil.

The Mid East Arabs understood the fifty-fifty deal, notwithstanding that Saudi Arabia refused to let the Venezuelans visit the kingdom because of their vote in the UN in favor of the partition of a Jewish state in Palestine. Saudi Arabia was the first Arab nation to promulgate a law taxing Aramco's revenues at a fifty-fifty split on December 26, 1950. In typical Mid Eastern fashion, they made the law retroactive to January 1, 1950.

Next year Gulf agreed to the fifty-fifty split with Kuwait; thus, the British AIOC had no choice but to go along or lose its 50% interest in the Kuwait Oil Company. Iraq soon followed with a fifty-fifty split with the Iraq Petroleum Company. The split brought the average nation's share to roughly 80¢ a barrel, so the all the other nations jumped on the bandwagon; and the oil companies had no choice but to cave into their demands.

However, there is always someone who doesn't get the big picture.

13

WHOOPS! — Somebody Goofed

Iran — The First Oil Crisis...Well, It Looked Like a Crisis

In 1949 Shah Muhammad Pahlavi went to Washington to seek economic aid and military assistance to defend his country from the Soviets. He was told Iran wouldn't stand a snowball's chance in the desert against the Soviet Union and, if he obtained a "fair deal from AIOC," he wouldn't need aid. For once, an Assistant Secretary of State knew what he was talking about — George McGhee was a wealthy oilman and aware of the fifty-fifty split in Venezuela and that Saudi Arabia was pressing for a similar deal.

Getting a fair deal from AIOC was easier said than done. AIOC had not lived up to paying the royalties under the 1933 agreement, which provided that Iran be paid per the value of gold rather than inflated British sterling; and the Iranians had yet to see AIOC's books. The Shah's new Prime Minister, General Ali Razmara, was a tough former Chief of Staff of the Army and reported to be honest. It was said that he refused bribes, something unheard of in the Mid East at the time. Razmara pressed AIOC for a fifty-fifty split, but Sir William Fraser, the tight-fisted Scot heading AIOC, wouldn't budge, even when prodded by George McGhee. McGhee gave up when the British government, which owned controlling interest in AIOC, backed Fraser.

Muhammad Mossadegh, the leader of a coalition of nationalistic political parties called the National Front, demanded the nationalization of AIOC. Razmara, aware of the power of the British government and lack of Iranian technical expertise to run the oil field, opposed nationalization and continued to work for a compromise. When Aramco and Saudi Arabia announced their fifty-fifty split, it was too late for Fraser to cave in. It was also too late for Razmara. He was assassinated in a mosque while praying, accused of being a "British stooge."

The Majlis passed a decree nationalizing AIOC and elected Mossadegh Prime Minister. At first, the British didn't believe the backward Iranians were serious or capable of running AIOC. They were right. The Iranians couldn't run the oil fields and refineries because AIOC had refused to train them. It wasn't until the Iranians showed up with a sign "Iranian National Oil Company" and sacrificed a sheep in front of the AIOC office that the British realized the natives were serious. (It was a temporary sign. The company became the National Iranian Oil Company.)

The British lodged a complaint with the International Court of Justice at the Hague (World Court), claiming the nationalization was illegal. This took a lot of chutzpah. Clement Atlee, Britain's Labour Party Prime Minister, had recently nationalized both domestic and foreign industries in Britain. Mossadegh wanted to continue producing and exporting oil while the World Court considered the case, but AIOC employees were told not to cooperate. In turn, "Old Mossy" as the British called him, twisted the British Lion's tail and ordered all AIOC employees out of Iran. A week later, AIOC employees marched aboard a British cruiser whistling the *Colonel Bogey March,*[50] leaving behind the largest oil refinery in the world.

The British responded with a unique form of embargo. They threatened nations and companies alike with law suits and that their oil supplies would be cut off if they purchased *their* Iranian oil. The Seven Sisters, with British help, convinced Truman that nationalization would spread through the Mid East if America didn't support AIOC. This opened the way for the American companies to obtain permission to join the embargo from the Justice Department, needed to avoid antitrust prosecution for collusion, with the argument a "tremendous world shortage" was imminent that would imperil the defense of the United States.

---

[50] *Colonel Bogey* is better known in America as the theme from the movie *The Bridge Over the River Kwai.* The English whistle it after they get their butts whipped.

It was declared the first international oil crisis, *but it wasn't.* All the Seven Sisters had to do was turn on the oil spigots in Venezuela, Saudi Arabia, Kuwait and Iraq. There was no shortage. The embargo worked. Things went to hell in Iran, which desperately needed the hard currency income from oil exports. Oil production in Iran plunged from 665,000 barrels a day in 1950 to 28,000 in 1952. During the same period, total world production soared from 11 to 13 million barrels a day, which made the Saudis, Kuwaitis and Iraqis happy. *(See Figure 2.)* Oil prices increased slightly, then dropped during the phantom shortage. Britain suffered because it had to pay, God forbid, American companies for the replacement oil and the British government lost profits from its 56% interest in AIOC.

Figure 2

OPEC FOUNDING MEMBERS OIL PRODUCTION
(000 bbl. day)

Nation	1950	1951	1952	Increase
Iran	665	350	28	(637)
Iraq	140	181	389	249
Kuwait	344	560	749	405
Saudi Arabia	547	762	825	278
Venezuela	1,498	1,705	1,804	306

Old Mossy became a worldwide hero or bum, depending on one's viewpoint, as he pled his case around the world and toyed with the Soviet Union. Actually, as a devout Muslim, he was anti-communist, but Americans saw communists behind every palm tree in Iran in those days. He was a cartoonist's dream. The bald, banana-nosed eccentric conducted the affairs of government in pajamas from his bed and was *Time* magazine's "Man of the Year" in 1951.[51]

It took two years for the World Court to rule in Iran's favor of its right to nationalize AIOC. As the case rested on basic international law, one has to wonder why it took two years to decide and was a ten to five decision, with the United States dissenting. So much for high-minded international judicial ethics. The judge from Great Britain was more honorable and true to the law and voted for Iran's position. However, the World Court's ruling didn't not stop the embargo and the stranglehold Britain put on the World Bank to prevent loans to Iran. Iran fell into economic chaos.

Britain, still suffering from the economic devastation of the war and a weak Labour government, returned Winston Churchill as Prime Minister. Churchill considered armed intervention and told Truman, "There might have been a splatter of musketry," and Britain would not have been "kicked out of Iran," if he had been Prime Minister. However, Truman was against using military force, and Churchill was unwilling to go it alone.

In the end, Mossadegh had to rely on the Tudeh communists and extreme nationalists to stay in power, and it appeared the Soviets were gaining influence. Islamic fundamentalists feared Mossadegh was becoming too powerful and had reduced the Shah to a helpless figurehead.

The time became ripe after Eisenhower's election and his appointment of no nonsense John Foster Dulles as Secretary of State and his kid brother, Allen W. Dulles, head of the Central Intelligence Agency. "Operation Ajax," a joint CIA and MI6 project to remove Mossadegh and restore the Shah to power went into effect. At first, it looked like the Monty Python plot was doomed to failure. The Shah ran off to Rome

---

[51] For those interested in Iran's early oil days, I suggest *Oil, Power & Principle: Iran's Oil Nationalization and its Aftermath* by Mostafa Elm. As it's dedicated to Old Mossy, you know where its bias lies, but it's well-written and far more factual than American and British versions.

with only the luggage he and his queen could carry. The State Department informed Eisenhower that Operation Ajax was a bust; however, the following day the Iranian Army rallied behind the Shah and arrested Mossadegh. The CIA and MI6 took the credit for the coup that would have occurred in the next few months anyway. In truth, the Seven Sisters did the real dirty work.

But that didn't mean AIOC won. It was plain that AIOC could not slip back into Iran and take over the oil fields and Abadan refinery. The Shah was not too keen on the idea either, as it would look like a British victory. The British were forced to split up the Iranian concession and hide AIOC in the pack. The State Department sent the perennial government oil consultant, Herbert Hoover, Jr., to help negotiate the deal with the Iranians and British, knowing he was anti-British. However, there were a lot of snags to overcome.

The Aramco partners, Exxon, Chevron, Texaco and Mobil, weren't happy about having to deal with the Iranians or AIOC. They had more than enough oil in Saudi Arabia and would have to reduce their take of Saudi oil in order to sell Iranian oil. Also, the Americans weren't too crazy about the political situation in Iran due to a weak Shah, nationalists who still wanted to run the country's oil business and religious fundamentalists who distrusted Westerners. Nor did Russia's eyeing of Iran and its oil make the companies feel comfortable. And there was something worse...

The Federal Trade Commission had discovered the Seven Sisters. In 1949 the FTC had prepared a report entitled *The International Petroleum Cartel.* The FTC's report read like an indictment of the oil companies for fixing oil prices. The State Department, Department of Defense and CIA jumped to the oil companies' defense, and Truman had it classified secret. It wasn't until 1952 that the report was released in "sanitized form." The FTC report hit the best seller list in world capitals. Readers in the Mid East, Venezuela and America were intrigued by the Red Line and Achnacarry "As Is" Agreements, and started to check the price of oil.

Public pressure forced Truman to order the Department of Justice to institute criminal antitrust proceedings against the Seven Sisters and the stepsister, CFP. The State and Defense Departments were against the action because they relied on the oil companies to "protect the national interest" and knew all hell would break loose if the facts came out. The British government, which owned controlling interest in AIOC and had been up to its ass in the oil companies' machinations, was pissed off. The French government, which owned 25% of CFP, told the United States to stick the criminal suit in a lower orifice.

One week before Eisenhower's inaugural, Truman ordered the Attorney General to drop the criminal action and prepare a civil suit. Eisenhower ordered Shell, AIOC and CFP dropped from the suit and limited the charges against the American companies to "a conspiracy to restrain interstate and foreign commerce of the United States in petroleum." The American Sisters received gentle slaps on the wrist by entering into consent decrees, which kept the dirty details out of the press by stating merely that they neither admitted or denied the accusations and promised never to do them again.

The State and Defense Departments paved the way for the Attorney General, with the blessing of the National Security Council, to allow the American oil companies to grab a hunk of the Iranian concession in the interest of national security without being in violation of the antitrust laws.[52]

Now, the Sisters could cut up the Iranian pie. AIOC ended up with 40% of its original concession although it had argued that it should have controlling interest. The American companies threatened to go home before they would allow the British to control the concession. Further, the other Sisters were not too happy about having to pay $90 million to AIOC for its interest plus a 10¢ per barrel royalty until $500 million had been paid. Shell walked off with 14% and CFP was doled out 6%. The Americans, Exxon, Chevron, Texaco, Mobil and Gulf, each

---

[52] If a reader is interested in the highly-slanted FTC and Department of Justice view, devoid of realism, read *The Control of Oil* by John M. Blair. For a moderate account, I highly recommend Daniel Yergin's Pulitzer Prize winner, *The Prize: The Epic Quest for Oil, Money and Power.*

ended up with 8%. However, as they had scads of oil in the Mid East and were still feeling the pressure of the antitrust suit, each gave 1% to a consortium of nine American independent oil companies called Iricon — tantamount to giving alms *(baksheesh)* to the poor.

Under the new agreement with the Shah, the oil companies were to manage Iran's oil industry and purchase its production; however, the National Iranian Oil Company would be recognized as owning the oil in the ground and the oil facilities...a first in the Mid East.

In the end... American oil companies controlled the majority of the oil in the Mid East. AIOC was so embarrassed, it changed its name to British Petroleum. Mossadegh went to jail for three years and died eight years later still under house arrest. Iran would be ruled by the egomaniacal Shah Muhammad Pahlavi until 1979.

Old Mossy had the last laugh from his grave after the Shah was deposed. Over one million Iranians would visit his grave on the anniversary of his death in 1979, *and the next Mid Eastern leader would be influenced by his nationalistic ideas.*

Egypt — The Second Oil Crisis, Well, Almost a Crisis

It started one Thursday when terrorists bombed a British base in Egypt. The British army, suspecting Egyptian police involvement, surrounded police headquarters in Ismailia and killed 50 Egyptians. The next day, January 26, 1952, was to be known as "Black Friday." Mobs, led by the Muslim Brotherhood, rioted and burned almost every hotel, store, theater, restaurant and bar associated with the British and Westerners in Cairo. Because Egyptian troops were not sent to quell the riots and stop the destruction, King Farouk was blamed and the Egyptian government fell into chaos.

On July 23 a group of Egyptian army officers calling themselves "Free Officers" forced King Farouk to abdicate. The successful coup was announced over Cairo radio by Anwar Sadat. Later, the Free Officers admitted that they were planning to boot Farouk out by the end of the year anyway, but decided there was no time like the present. Farouk went into exile on the French Riviera with a lot of bucks he managed to stash away, like most kings do, where he became a famous playboy and renown for being seen with beautiful bimbos and being as fat as a pregnant buffalo.

The Free Officers organized into the Revolutionary Command Council with General Muhammad Neguib as President and Lt. Colonel Gamal Abdul Nasser as second in command. For a brief time, no one knew what the Free Officers were up to other than they promised land reform to please the peasants and announced they were anti-British, which pleased everyone. Establishing a democracy was not on their agenda. They dissolved all political parties when they heard of communist opposition and the Muslim Brotherhood attempted to assassinate Nasser in October 1954. Not trusting old General Neguib, Nasser ousted him from power and placed him under house arrest. [53]

A fervent nationalist, Nasser was the first Egyptian to rule Egypt since the time of the ancient Pharaohs. (Remember, Cleopatra was a Greek.) One of the first things he did was negotiate a treaty with the British to remove all their troops by June 1956. Then he took advantage of the Cold War and set about playing off the United States and Britain against the Soviets. When questioned about communism, Nasser replied that he was more concerned about having Israel for a neighbor. (In the 1950s, all successful Arab leaders had to hate the Jews.)

Nasser's greatest dream was the construction of the $1.5 billion Aswan Dam to harness the waters of the Nile for agriculture and electricity. The dam promised energy for Egyptian industry and fertile land for 10 million peasant farmers — an Egyptian Tennessee Valley Authority government boondoggle. The United States thought it was an ideal project to underwrite at the World Bank with a loan of $200 million (an enormous amount in those days) and seal its friendship with Nasser. However, the United States refused to supply Nasser with arms, for fear it would allow Egypt to upset the balance of power against

[53] General Neguib and Mossadegh were both put under house arrest. It is a common practice in the Mid East and a way of telling the person arrested that it is politically incorrect to execute you or keep you in jail but, if you leave the house, we'll shoot you.

Israel, forcing him to enter into a cotton-for-arms deal with the Soviets and invite more Russian "advisors" to Egypt than America thought necessary. Then Nasser had the audacity to recognize Red China, a definite *no, no* as far as American foreign policy was concerned, which could be summarized "better dead than red."

On July 19, 1956, Secretary of State Dulles announced the cancellation of the funding for the Aswan Dam.[54] One week later, in celebration of the fourth anniversary of the Egyptian revolution and with no British troops to stop him, Nasser nationalized the Suez Canal, in which two-thirds of Europe's oil had to pass or go around the southern tip of Africa. Diplomatic negotiations failed to resolve the issue. Nasser refused to recognize the 1888 Convention of Constantinople, guaranteeing all nations the right of transit through the Suez Canal, labeling it a European colonialists' scheme to deprive Egypt of its sovereignty.

Nasser had taken a stand and couldn't back down without losing face in Egypt and the Arab world. Nasser's position on nationalization of the Suez Canal was similar to Mossadegh's on Iranian oil — the Suez Canal was a natural resource of Egypt, but the bulk of its income was going to the British government, the Canal's largest shareholder, as did the British government's shareholder interest in AIOC. The British and French pilots who steered the ships through the narrow canal were ordered out of Egypt by their governments to put pressure on Nasser. To the West's surprise, the Egyptians, with a little help from the Russians, managed to keep traffic flowing with far less confusion and problems than expected. (So, a couple of boats ran into each other.)

Despite warnings from Eisenhower, the French and British secretly planned armed intervention to protect *their* Suez Canal. Aware Britain had a history of sending troops to the Mid East when it didn't get its way,

---

[54] Dulles was an aggressive Secretary of State known for his tough stance against Soviet Russia and policy of "brinkmanship" — daring to take a hazardous course of action to the brink of catastrophe. Pilots landing at Dulles International Airport in Virginia often think of the man for which it was named when descending in the fog.

Syria and Jordan placed their armed forces under Egypt to protect the Suez Canal. On October 24 French, British and Israeli leaders met in Sévres, France, to concoct a cockamamie plan calling for Israel to attack across the Sinai Desert in response to Egyptian threats to cut off their shipping. The British and French would then attack to protect the Suez Canal as an international waterway. At the time, the Soviets were embroiled in crushing an anti-communist uprising in Budapest, and the three plotters thought the Russians would be too busy in Hungary to help their buddy, Nasser.

On October 29 Israeli tanks started to roll across the Sinai. As planned, on October 31, the British bombed Egyptian airfields, crippling the pathetic Egyptian Air Force. The Syrians and Jordanians wet their pants when they heard about the might of the RAF and Israeli army in the desert. But there was a SNAFU in the logistics. The British and French paratroopers didn't land in the Canal Zone until November 5.

The aggressors forgot an important date. President Eisenhower, who had told the American public the United States should remain neutral in the Suez crisis and spoke out strongly against the use of force, was mad as hell. He first heard about the attack on Egypt while campaigning for reelection in Richmond, Virginia. **November 6, 1956, was election day in the United States.** Ike told Britain, France and Israel to get the hell out of the Canal Zone.

The Anglo-French force agreed to a United Nations-ordered cease fire on November 7. By that time, Egypt had scuttled ships in the Suez Canal filled with rocks, cement and, believe it or not, empty beer bottles, blocking passage. The Syrians blew up the pipelines from Iraq and Saudi Arabia to the Mediterranean through Syria; and Saudi Arabia declared an oil embargo against Britain and France. During the confusion, Russia rattled its nuclear saber, which really teed off Ike. He told Nikita Khrushchev that he would respond, "as sure as night follows day," which shut up the fat little Russki.

With winter approaching it appeared Europe would be without oil. There were not enough tankers available to make the long expensive journey around Africa. Britain turned to the American oil companies. Although Ike won reelection in a landslide, he blamed the three invaders for the loss of several key Republican Congressional seats. Ike told the Brits to boil in their own oil and refused to allow American oil firms to operate in concert and grant them protection from United States antitrust laws until all three nations withdrew from Egypt. When they finally pulled out, tankers were rerouted and plied between America and Europe, which could be done twice as fast as from the Persian Gulf to Europe via the Cape of Good Hope.

Not all the American oil companies cooperated with the plan. The Texas Railroad Commission, controlled by Texas independent oil companies, refused to allow any increases in production under the bugbear market demand prorationing — there was no oil shortage in America! It wasn't until Exxon raised the price of oil 35¢ a barrel did the TRC allow the Texans to produce at a full capacity.

In the spring of 1957, the pipelines to the Mediterranean resumed flowing and the Suez Canal was cleared for passage. The second so-called oil crisis was over. There was never an actual shortage of oil, only a scarcity of available tanker transportation to the European market. The Suez Canal, long regarded by Europe as a secure route, could no longer be regarded as safe. Also, pipelines shortcutting the distance and cutting costs could easily be sabotaged by a few well-placed explosives. The loss of the Suez Canal added 4,500 miles to the already long 6,500 mile journey. Thus arrived the age of supertankers — giants four football fields long capable of carrying three million barrels of oil.

Nasser was the clear victor. He had rid his country of the British colonialists not once, but twice. Arab nationalism and anti-Western feelings boomed in the Mid East, with Nasser as the leader and hero of the Arabs. Despite Eisenhower's support in the Suez, which Nasser could not have survived without, Nasser became more reliant on the Soviets, who promised to assist in the completion of the Aswan Dam. If John Foster Dulles had not been stricken with cancer a few days before the crisis, the outcome may have been different. From his hospital bed, he told Eisenhower and the British Foreign Secretary that the British should have marched to Cairo and gotten rid of that SOB Nasser.

Eisenhower proclaimed what was to be called the Eisenhower Doctrine by pledging military and economic aid to Mid East nations opposing communism. Although similar to the Truman Doctrine, the Eisenhower Doctrine was a flop. The Arabs were more interested in getting rid of Israel and America wouldn't supply bigger guns to the Arabs than they were sending to Israel. Ike kept a wary eye on Nasser while aligning the United States with Saudi Arabia, the most stable nation in the area and the one with the most oil. America's position during the Suez crisis infuriated the British, French and Israelis, but was welcomed by many Arabs, particularly Saudi Arabia. All over the Mid East acts of sabotage had slowed the delivery of oil to the West except in the land of Aramco — Saudi Arabia.

The Big loser was Britain. The loss of face in Egypt signified its downfall as a colonial power throughout the world. The British Lion's last roar was a meow.[55]

IKE'S HIGHWAYS

In 1919 Army Captain Dwight D. Eisenhower led a military convoy from Washington, D.C. to San Francisco to show the need for highways across the nation. It took 63 days to travel the muddy roads and thin asphalt that could not bear the weight of the heavy army vehicles. In 1956 President Eisenhower signed the Interstate Highway Act, providing over 40,000 miles of superhighways crisscrossing America.

The interstate system gave a boost to America's love of the automobile, added to the demand for gasoline *and increased the country's need for imported oil.*

---

[55] Chester L. Cooper, *The Lion's Last Roar: Suez, 1956.*

14

THE GAS GUZZLING FIFTIES — News Flashes & Obituaries

No history text can cover every event that took place in all the nations of the Mid East during the turmoil of the 1950s through today. Indeed, historians record only what they believe to be the significant events affecting a civilization or nation and enjoy debating which were the most meaningful. Primarily, they write of wars and conquests that resulted in the expansion or disappearance of a nation or peoples. Historians trace the development of peoples through their architecture, art, literature, laws, agriculture, technology and religion, which is beyond the scope of this primer, except to point out the great significance of Islam and its opposition to Judaism that has its roots in the Koran.[56]

Too often, economics is overlooked or given a back seat in history. In the modern Mid East, religion and economics are in the forefront of history. Oil, today's greatest economic prize, changed the history of the Mid East. Without oil there would have been little for the Western world to covet and fight over and Mid Eastern history probably would have ended with the fall of the Ottoman Empire except for the carving up of Palestine. It is not unfair to say that few would have heard of Kuwait;

---

[56] In *suras* (chapters) 61 and 62 of the Koran, the Israelites are chastised for denying Jesus and extinguishing "the light of God;" and that to entrust the Torah to the Jews was likened to "a donkey laden with books."

and it's a 100 to one bet that the average American, Joe Sixpack, would not have heard of the tiny United Arab Emirates (UAE) unless the nation had almost *five times* as much oil as the United States...Many people don't know the UAE exists today.

Hence, I will concentrate on the unstable Mid East politics, coups, wars and economics — oil and its progeny, the Organization of Petroleum Exporting Countries.

To complete the 1950s and step into the 1960s it's necessary to scan the following *NEWS FLASHES:*

SAUDI TROOPS INVADE BURAIMI OASIS 140 miles inside territory also claimed by Oman and Abu Dhabi in August 1952. British-led Omani scouts drive Saudis out. CIA involved in attempting to gain additional oil lands for Aramco concession. [The claim was settled in World Court with Abu Dhabi gaining the lands after alleged CIA attempts to bribe arbitrators. They never found oil in the area.]

BAGHDAD PACT SIGNED. In February 1955 Iraq and Turkey sign a mutual defense treaty against USSR aggression. Britain, Iran and Pakistan join. Nasser says Iraq betrayed the cause of Arab unity and seeks ties with Saudi Arabia and Syria. [Much ado was made about the pact, but it never amounted to anything. Iraq pulled out in 1958 after a coup and the pact changed its name to Central Treaty Organization.]

EGYPT & SYRIA UNITE — FORM UNITED ARAB REPUBLIC on February 1, 1958. After a series of coups which resulted in Syria having six presidents since 1949, the Syrians throw up their hands and ask Nasser to run their unruly rabble.

IRAQ & JORDAN FORM ARAB UNION on February 14, 1958, but they never get around to agreeing how to unify the nations except to adopt the red, green, black and white flag of the Arab revolt designed by Sir Mark Sykes in 1916.

Editorial

Most Americans never heard of the Arab Union, but it bent a lot of Iraqis, Jordanians and their Arab neighbors out of shape.

PRINCE FAISAL TAKES OVER IN SAUDI ARABIA from his brother, King Saud, Ibn Saud's successor, in a quiet intrafamily coup in April 1958. Since taking over after Ibn Saud's death in 1953, King Saud had been accused of corruption, decadence and bankrolling a failed assassination attempt of Nasser.

MONARCHY OVERTHROWN IN IRAQ on July 14, 1958. Brigadier Abdel-Karim Kassem takes over in a bloody coup. King Faisel II, the Crown Prince and Nuri al-Said are killed and their mutilated bodies are dragged through the streets. Kassem, a brutal psychopath on a par with his junior aide, Saddam Hussein, declared Iraq's support of Nasser, which meant he was up to no good as far as America and Britain were concerned.

EISENHOWER ORDERS 10,000 MARINES TO LEBANON a week after the Iraqi coup at the request of Christian President Camille Chammoun of Lebanon to prevent his overthrow by Muslim militias.

   British Send Troops to Protect King Hussein of Jordan, who feared a revolt and was already regretting removing Glubb Pasha as the head of his British-trained Army in order to placate the pro-Nasser nationalists and Palestinian refugees.

Editorial

Why is the United States involved in Lebanon? Nasser's Arab nationalism is spreading through the Mid East like a forest fire and Lebanon and Jordan are flooded with Muslim Palestinian refugees. The Christian pro-western President wants to change the constitution and give himself a second term in office. The Muslims say "no way." Many Christians don't want him for a second term, either. Lebanon is in the midst of religious and political turmoil. When the French created Lebanon, it added only enough Muslims from Greater Syria to give the Christians a slight Christian majority. Forgetting that the Muslims might breed faster than Christians, the French set up a "Confessional system," providing for a Maronite Christian President, Sunni Muslim Prime Minister and Shiite Speaker of the legislature.

A second reason for American Marines being sent to Lebanon was Nasser's mumbling that Eisenhower was "all mouth" and wouldn't act, so Ike set the record straight.

IRAQ'S ABDUL KASSEM SURVIVES COUP & ASSASSINATION ATTEMPT in October 1959. Kassem professed Nasser's ideals, but he wasn't about to join the United Arab Republic and become second banana to Nasser. With the support of the Communist Party and Kurds (whom he later slaughtered), Kassem eliminated all opposition, including the "Free Officers" who helped him to power. (Few in the press reported that one of the wounded attempted assassins who escaped to Egypt was the twenty-two year old weirdo, Saddam Hussein.)

AL-FATAH FORMED BY YASSER ARAFAT in 1958 in Kuwait, which was inundated with 200,000 Palestinian refugees. Members of the Arab League began financial support for the Fatah and it public offspring, the Palestinian Liberation Organization (PLO), and other hit squads of the *fedayeen* (those who sacrifice themselves) eventually to be led by Arafat.

Cartoonists were unaware that baggy pants, scruffy Arafat's real name was Muhammad Abdul Rauf Arafat al Qudwa and his buddies nicknamed him Yasser, which means "carefree."

FIFTY-FIFTY SPLIT BROKEN in 1957. Only the oil and gas trade journals reported the news that Enrico Mattei of Italy breached the fifty-fifty profit split held sacred by the Seven Sisters. Fed up with attempting to finagle Italy a share of Mid East oil from the Seven Sisters, Mattei made an offer the Shah couldn't refuse — a split of 75% for Iran and 25% for the Italian company, Ente Nazionale Idrocarburi (ENI). At first, it looked like Mafioso blackmail. Mattei offered to back out of the deal if Aramco gave ENI 10% of its Saudi concession and the Sisters surrendered 5% of the Iranian consortium. The Sisters told the [censored] Italian *Vada Via!.* Like the Italian navigator who couldn't find Bahrain and dropped bombs in Saudi Arabia at the beginning of World War II, ENI couldn't find oil, so the concession was declared *finito.*

The Japanese negotiated a 56-44% split with Saudi Arabia for a concession in the offshore area of the Neutral Zone, then a 57-43% split with Kuwait. Hearing the Kuwaitis got a better deal, King Saud told the Japanese to cough up another one percent. After the Japanese hit oil, Saudi Arabia and Kuwait grabbed another 10% from the Japanese who were neophytes in dealing with the Arabs, breaking the fifty-fifty split even wider.

Standard Oil of Indiana (Amoco), the largest American oil company not a member of the Seven Sisters and desperate to hold its position in the United States, made shambles of the split by agreeing to the Italian split of 75-25% with the Shah in 1958 and struck a vast oil field offshore in the Persian Gulf.

Editorial

The Mid Eastern nations and Venezuela didn't have to be economists or CPAs to figure out that if the Japanese and Amoco could afford to pay such high royalties and taxes, so could the Seven Sisters.

OIL DISCOVERED IN LIBYA IN 1959. The world had a surplus of oil, thus, the gusher of sweet North African crude oil that could be made into a high percentage of gasoline and clean oils went unnoticed by the public, but not the Seven Sisters, eight if you count the French stepsister who had discovered oil in Algeria in 1956. Under old King Idris, Libya was considered the most stable North African nation. Algeria was engaged in a war for independence from France. Libyan oil also had the advantage of cheaper transportation costs because of its close proximity to Europe and it was not susceptible to a Suez Canal blockade.

Wise King Idris and his oil minister, aware of the control the Seven Sisters had over Mid East oil, insisted Libya not be burdened with a single oil concession. Idris awarded concessions to seventeen oil companies, mostly independents, who were not caught up in the web of balancing the production of various Mid Eastern nations, such as Saudi Arabia and Iran, both of whom were clamoring for more royalties. The Libyans knew that the oil-hungry independents would turn the oil spigot wide open.

The World is Flooded With Cheap Oil, Except for America

When cheap Mid Eastern and Venezuelan oil started to flow into the United States in the mid 1950s and cut prices, American independent oilmen cried foul. Big Oil's profits from domestic production also dropped, but their ability to import cheap crude oil sustained their overall profits and maintained their U.S. oil reserves. Complaints were raised by the Senate Majority Leader, Lyndon Johnson, and the Speaker of the House, Sam Rayburn — both Texans — that the American oil industry would be decimated and America's national defense was in jeopardy. On the other side, conservationists voiced concern that American oil reserves were being depleted, but they would lose their arguments to the politics of oil.

To the shock of the most advanced nation in the world, America discovered it was dependent on Third World countries for its oil. Portraits of rich Arabs driving Rolls Royces and boozing in Paris nightclubs with beautiful blondes appeared in the tabloids and rankled the public. Venezuela, America's biggest supplier was no longer our "Good Neighbor" after an angry mob of "communists and leftists" attacked Vice President Nixon's motorcade while on a good will visit to Caracas in 1958.

Congressmen from oil producing states demanded an increase in the oil tariff from 101/2¢ to $1.05 a barrel. However, Eisenhower opposed tariffs and asked the oil industry to voluntarily limit the amount of imports in 1957. Importers ignored the voluntary quotas or could not comply because the Mid Eastern nations were demanding greater production — they wanted the cash — and making noises that their concessions were in jeopardy if they didn't increase their take. In 1959 Eisenhower imposed the Mandatory Oil Import Program limiting the amount of imported oil to nine percent of the nation's demand based on Chicken Little cries of "national security" to protect the domestic oil industry from cheap foreign crude oil.

NATIONAL SECURITY

In the interest of national security, only oil imported overland was exempt, which allowed only Mexican and Canadian oil to be imported, but Mexico had no pipelines to the United States. This invited a scam called the *Mexican Merry-Go-Round*. Oil was shipped to Brownsville, Texas, loaded into trucks and driven to Mexico, where it turned around and drove back to Texas.

To assure small refiners remained competitive with major refiners importing foreign crude oil, importers were required to purchase "tickets" from small refiners without access to imported oil to offset the lower price advantage of imported oil, which eventually reached a value of 45¢ a barrel. The government boondoggle subsidized the small refiners and maintained big oil's profits from the pockets of the consumer.

Crude oil prices leveled off at $2.90 a barrel in the United States and $1.80 in the Middle East.

QUIZ: What is meant by the "price of oil?"

Answer: Crude oils vary in physical character, such as sweet or sour (sweet less than 0.5% sulfur and sour over 1.5% sulfur), gravity and impurities. The ability to refine a particular crude oil into the value of its finished products determines its price. As there are many types of crude oil, quoted prices refer to "marker crudes." In the United States, the benchmark crude is West Texas Intermediate. In the Mid East, the marker crude oil was Saudi Arabian Light before it was changed to a "basket" (average) of eight crudes. When sold, price adjustments are made for each crude based on its value compared to the marker crude oil.

The above definition of the "price of oil" sounds too simple. What else was involved?

Answer: Lots, including the *transportation differential* (tanker costs); however, the following three issues were at the forefront of the international oil companies' concerns in the 1960s: **First:** the *posted price* the companies announced they were charging and the *actual price* charged. Discounts off the posted price were often given in times of low demand and premiums added in times of high demand or shortage. **Second:** the applicable tax and royalty formula and how it was to be paid. In most cases (Libya was an exception), it was based on the posted price, which meant that in times of low demand when discounts were allowed, the oil companies' profits were cut. **Third:** the production costs. This was important to Venezuela where production costs were 80¢ a barrel compared to between 10¢ to 20¢ in the Mid East.

All of these issues were about to raise their ugly heads in the coming years.[57]

---

[57] For an excellent overview of petroleum economics, I recommend *Where's the Shortage? A Nontechnical Guide to Petroleum Economics* by Bob Tippee. It is must reading for those not familiar with the oil industry who desire to avoid being flimflammed by the mournful cries of Big Oil, campaigning politicians and whining do-gooders or consumer advocates.

15

THE BIRTH OF OPEC — Almost an Abortion

Contrary to conventional belief, the father of OPEC was not an Arab, but a Venezuelan, Juan Pablo Pérez Alfonzo. After the overthrow of the corrupt dictator Colonel Marcos Pérez Jiménez in 1958, Pérez Alfonzo returned from exile for the third time and was appointed the Minister of Mines and Hydrocarbons by the new President, Romulo Betancourt. During his last banishment, he lived in Washington, D.C. where he studied the United States petroleum industry and became fascinated with the Texas Railroad Commission and market demand prorationing in order to maintain higher oil prices.

In 1959 the United States Mandatory Oil Import Program drastically curtailed Venezuela's oil production, where 40% of its exports were shipped. Pérez Alfonzo was given the finger when he visited Washington to plead for a special quota for his nation. America distrusted the left wing government of Venezuela and was angered by the attack on Vice President Nixon during his visit to Caracas the previous year. He objected to the preference given Mexico merely because the oil could be transported by land, a joke — there were no pipelines from Mexico. His arguments fell on deaf ears that it was Venezuela, not Mexico, that supplied oil to the United States during World War II; and it was Mexico, not Venezuela, that nationalized the American oil companies in 1938.

Pérez Alfonzo's next stop was Cairo to attend the Arab League sponsored First Arab Oil Congress in April 1959. As an observer, he went to make friends with Venezuela's competitors, aware that most of Venezuela's crude oil was heavier and more sour than the majority of Mid Eastern crudes, hence, less valuable, and its 80¢ a barrel production costs far exceeded the 20¢ average in the Mid East by more than enough to make it uncompetitive. His aim was to introduce market demand pro-rationing à la the Texas Railroad Commission quota system, which would also raise the per barrel cost of Mid East production. As part of his spiel, he planned to push conservation of a wasting asset, just as Big Oil had done in the East Texas field in the 1930s, in hope of getting the conservationists on his side.

At the start, the meeting looked like it might be a bust. Of the 400 delegates, three quarters were from Syria and Egypt, which had little oil and were there to "show the masses the importance of oil" and "coordinate the efforts of Arab Governments." "Masses" was a pinko-commie phrase to some attendees, and "Arab" meant that Iran was not invited. Nevertheless, Iran sent an observer to see what the Arabs were up to. Iraq refused to send a delegation because it was being held in Egypt. Iraq's Abdul Kassem believed that Nasser shouldn't have anything to do with Iraqi oil. Muhammad Salman, an Iraqi who was an Arab League official sponsoring the shindig, had to put in an appearance. A representative from Kuwait, then the largest Mid East producer, sat in on the meeting but said little because Kuwait was still a British protectorate.

It was Abdullah Tariki of Saudi Arabia who cemented a relationship with the Venezuelan. Since 1955 Tariki had been Saudi Arabia's unlikely Director of Oil and Mining Affairs by default. The "Red Sheik" was one of the few Saudi graduates in chemistry and geology. After obtaining a degree from the University of Texas, Tariki worked for Texaco in Texas as a trainee geologist. Some said he hated Americans because he was thrown out of bars on numerous occasions by Texans who thought he was a Mexican. Tariki's socialistic ideas differed from Pérez Alfonzo's in that he wanted to nationalize Aramco, then learn how to run the company.

Rather than debate their ideas before the entire Arab Oil Congress, most of whom knew little about oil other than it earned a lot of money that they were anxious to get their hands on, the key participants adjourned to a local yacht club in the suburbs of Cairo. There they signed a general agreement called the Maadi Pact, endorsing the concepts of a 60-40% split in favor of the governments, company price changes should be discussed with the governments, government involvement in refining and downstream operations, and the establishment of national oil companies to control conservation and production. However, it was clear that their biggest gripe was the oil companies should not change the price of crude oil, at least downward, without consulting with the governments.

Of interest, the Maadi Pact was signed by the largest oil producer, Kuwait, which insisted on no notoriety because it was still a British protectorate and wasn't supposed to be talking to other nations about something as important as oil; Saudi Arabia and Venezuela by two left leaning connivers with bigger things on their minds; the United Arab Republic, which had little oil; and Iran and Iraq as observers.

Tariki and Pérez Alfonzo were so elated with their progress, they went to Texas to attend the annual meeting of the Texas Independent Producers and Royalty Owners. The Texans were bitching about foreign oil imports and were glad to hear that there was a move afoot to limit Mid East production and raise prices. From there, Pérez Alfonzo took Tariki home with him to Venezuela, where they pinned a medal on the pinko Red Sheik.

The Maadi Pact was kept secret, although there were demands on the companies by the Venezuelan and Mid East governments to consult with them before changing prices. On August 9, 1960, oil and lots of other smelly stuff hit the fan. Without any consultation, Exxon reduced the posted price of Mid East oil by 5¢ to 14¢ a barrel. Saudi Arabian Light was cut from $1.90 to $1.76, or 7%. Tariki and Pérez Alfonzo screamed bloody hell in public, but smiled behind closed doors with the knowledge they could now get the Maadi group organized and put their plan into effect.

The six other Sisters plus the French stepsister, CFP, were quick to point their fingers at Exxon's foolhardiness, but swiftly fell in line with similar price cuts. Pérez Alfonzo warned Exxon and Shell not to drop their prices in Venezuela — "or else." Shell and Exxon got the message, but managed to reduce the amount of Venezuelan exports to cut their losses.

Exxon's excuse for cutting prices was valid, but its brashness lame-brained. The Mandatory Oil Import Program effectively denied Mid

Eastern oil and severely limited Venezuelan oil imports into the United States. Europe was being flooded with cheap oil from Russia, which had recently reentered the market and was selling oil at bargain basement prices (one-half the Mid East prices) because of its dire need of hard currency. The independent oil companies were adding to the glut and not hesitant to sell at any price. During the past year, the Sisters had offered discounts off the posted price to keep the Mid East governments happy and were feeling the squeeze on profits because they had to pay royalties and taxes on the higher posted price.[58]

The Maadi Pact group decided to meet...but where? Cairo was out. Iraq wouldn't attend if Nasser had anything to do with the group. The UAR had broken off diplomatic relations with Iran. Besides, the UAR was only a marginal oil producer. As a result, the meeting was held in Baghdad and Nasser's UAR wasn't invited.

The invitees were Saudi Arabia, Kuwait, Iran, Iraq and Venezuela. Qatar showed up but was not allowed to join because it was too small and would make the new club appear "too Arab." The three Arab members put aside their differences with Iran for supplying oil to Israel, which they were boycotting, and selected the Iranian delegate, Fuad Rouhani, to chair the meeting. The five attendees who formed the *ORGANIZATION OF PETROLEUM EXPORTING COUNTRIES (OPEC)*, the name proposed by Pérez Alfonzo, owned over 80% of the world's crude oil exports.

---

[58] Many have written about the events leading up to the birth of OPEC with the benefit of 20/20 hindsight. I have already suggested Daniel Yergin's *The Prize* as a well-documented account of the times. In addition, I highly recommend *OPEC: Twenty-five Years of Prices and Politics* by Ian Skeet for a thorough analysis that leads a reader through the complex political and economic entanglements with skill and clarity. I do not recommend *OPEC: The Inside Story* by Pierre Terzian unless you are fond of left wing drivel and don't mind wading through a translation from French that appears to have been done by a freshman from remedial English class whose first language was Swahili.

Iran's Rouhani did a remarkable job, in part, because he had no specific instructions from Shah Muhammad Pahlavi other than not to agree to anything stupid or violent. Rouhani managed to keep Tariki from demanding the new organization immediately tax the companies at a rate of 60% and Pérez Alfonzo from cutting production to raise prices.[59] The Iraqi delegate didn't know very much about oil and contributed little; and the Kuwaiti delegate sat quietly as Kuwait was still a British protectorate.

The substantive OPEC resolutions at its creation on September 14, 1960, appeared reasonable. In effect, the members said: (1) They could no longer remain indifferent to the attitude of the oil companies in effecting price changes. (2) Prices should remain free from unnecessary fluctuations and restored to the pre-August 1960 level; and that in case of price changes the companies "shall enter into consultation with the Member or Members affected to explain the circumstances." (3) The members should study and formulate a system to stabilize prices through the regulation of production, with due regard to the interests of the producing and consuming nations in order to secure a steady income to the producing nations and a fair return on capital to the companies. (4) If as a result of any OPEC unanimous decision, sanctions are employed by the companies, the members would not break ranks and accept benefits to the detriment of another member.

The Western press gave OPEC scant attention. Moscow's *Ekonomicheskaya Gazeta* commented: "The establishment of such an organization is a new feature in the struggle of the peoples of economically underdeveloped countries against the domination of monopoly capital." Most oil companies yawned and predicted that OPEC wouldn't work.

The members had not asked for much. The companies had long recognized that the nations could not remain indifferent to price fluctu-

---

[59] Pérez Alfonzo arrived late at the meeting due to an attempted coup in Venezuela. Noteworthy was Pérez Alfonzo's failure to mention private chat about the meeting with the USSR a week earlier.

ations, as their economies were dependent on oil revenues. Big Oil was also striving for stable prices and greater profits. The commitment to study a system of regulation to stabilize prices was merely a bone to Pérez Alfonzo and admitted to a *fair return on capital,* whatever that meant. The companies recognized they had to *consult* with the governments before changing posted prices, which formed the basis of their national budgets. In practice, the companies adhered to the demand because they could lower *actual prices* and the near future faced only increases in posted prices.

The last resolution was a concept of Rouhani who was well aware of the Seven Sisters' boycott of Iranian oil after the nationalization of AIOC in 1951. Theoretically, the Sisters would not be able to shut in one member nation and turn on the oil spigot in other member nations to make up the difference.

But there was a catch. OPEC actions required unanimous decisions by its members...The Seven Sisters laughed...How could three backward Arab nations, a self-centered Shah who insisted on running everything and a banana republic ever agree on anything?

ARAB PROVERB: "If the sailors are too numerous, the ship sinks," was heard throughout the Arab world. Americans say it differently: "Too many cooks spoil the broth."

Getting Organized and Finding We Don't Agree

The second OPEC conference was held in Caracas in January 1961. At first, things went smoothly. Rouhani was confirmed as the Secretary General and little Qatar was admitted as a member, not that anyone noticed. They divided the Secretariat into four departments — public relations, administration, technical and enforcement, although they didn't have anything to administer or enforce. It was apparent the members knew little about the technical and economic aspects of the oil industry. An American consulting firm was hired to perform a study of petroleum economics. OPEC would not get around to establishing an economics department until 1965 when someone mentioned the law of supply and demand.

Geneva, Switzerland, was chosen as a neutral ground for its headquarters because of the members' practice of boycotting meetings held in countries they were squabbling with. Later, the headquarters had to be moved to Vienna, Austria, because the Swiss didn't think OPEC was important enough to grant diplomatic status.

After getting organized, the members discovered they couldn't agree on anything important. Each nation was diverse and had its own problems and political axes to grind.

Venezuela wanted to reduce and prorate oil production as a means of maintaining prices and competing with its new friends from the Mid East.

Iran, under the egotistical Shah, was hellbent to regain its status as the number one Mid East producer it held prior to 1951 and build a big army. He said prorationing was a pipedream ..."the Arabs would cheat." (He was right.)

Kuwait said little — it was still waiting for its independence from Britain in June 1961. Its National Assembly, under pressure from 250,000 Palestinian refugees, didn't support anything OPEC did under the mistaken belief they had a voice in government. When the legislature got out of line, Sheik al-Sabah dissolved it.

Figure 3
OPEC FOUNDING MEMBERS
OIL PRODUCTION
(000 bbl. per day)

Nation	1950	1960	1970
Iran	665	1,067	3,829
Iraq	140	972	1,549
Kuwait	344	1,692	2,990
Saudi Arabia	547	1,314	3,799
Venezuela	1,498	2,846	3,708

Petroleum Economist

Iraq was also waiting for tiny Kuwait's independence. Abdul Kassem moved to claim Kuwait as part of its Basra province when the British left in 1961, as would Saddam Hussein 30 years later. British troops, supported by Saudi, Jordanian and the UAR forces, forced Iraq's retreat. Iraq boycotted OPEC until after Kassem was assassinated in 1963. In 1961 Kassem took matters in his own hands and confiscated all Iraq Petroleum Company concession lands (99.5%) not being exploited.

Saudi Arabia, concerned about the rise in power of the Shah and Nasser, was not about to reduce its oil revenues or surrender power in the area.

The disparities between the OPEC member nations, particularly Saudi Arabia and Iran ran deeper than Arab versus Aryan. The Saudis were primarily Sunni Muslim and the Iranians were the more fundamentalist Shiites, a major source of discontent in Saudi Arabia. When the companies failed to produce enough oil to his liking, the Shah played footsy with Russia until the United States and Britain used "moral suasion" — twisted the arms — of the oil companies to increase Iranian production. As a kingdom, Saudi Arabia wasn't interested in democracy and feared communism, thus, couldn't take advantage of the Cold War. The Shah's tactics worked — Iran nosed out Saudi Arabia as the top Mid East producer in 1970. *(See Figure 3.)*

Until 1967 OPEC would continue to meet and accomplish little. Only Libya and Indonesia thought the club was worth joining in 1962. Arab Libya seemed a likely member because of its mounting oil production, but Indonesia in the distant Pacific served an entirely different market in the Far East and wasn't burdened with the same price pricing issues. However, the fact that Indonesia was 90% Muslim raised a few eyebrows.

The simple reality was there was too much oil available in the world and import quotas kept foreign oil out of the United States, which guzzled one-third of the world's oil. OPEC's demand to return to the pre-August 1960 prices faded and Pérez Alfonzo's quota system was unacceptable to nations in dire need of revenues.

QUIZ: What happened to the two pinko dirt bags who organized OPEC?

Answer: The "Father of OPEC," Pérez Alfonzo, came to the conclusion that OPEC was run by greedy sheiks and a shah, and Venezuelan politicians weren't much better. He resigned as Venezuela's petroleum minister in 1963, writing off OPEC as the savior of the masses and labeling oil "the excrement of the devil," which is another way of saying it was really bad shit.

In 1962 Tariki made the mistake of accusing King Faisal's brother-in-law of taking *baksheesh* on the Japanese oil concession and was fired and sent into exile. Taking his place was a thirty-two year old lawyer who would become known as "Mr. OPEC" and the symbol of Arab oil — Sheik Ahmed Zaki Yamani.

Under the theory that every organization needs a cause to rally behind, OPEC finally agreed on a boring accounting issue — royalty expensing — which boiled down to whether the government tax should be computed before or after the companies deducted the royalty. The difference amounted to approximately 11¢ a barrel.

Rouhani, as Secretary General, discovered that Big Oil played hardball. The first fast ball came low and inside. The companies ignored OPEC. OPEC was not a signatory to the contracts with the various governments. In other words, they said Rouhani could only represent Iran as an official of the National Iranian Oil Company, and the company officials would only represent their companies, such as Aramco, Iraq Petroleum Company and the Iranian consortium that replaced AIOC. After three years of negotiations, the Shah subverted OPEC's unity and accepted a lower company proposal. Rouhani was left with egg on his embarrassed face. He couldn't represent OPEC because the Shah had made an end run OPEC's negotiations.

QUIZ: Why quibble over 11¢?

Simple answer: It's money...lots of money.

Between 1960 and 1970 the posted price stayed relatively constant at around $1.80 a barrel. The average government's share was 80¢; thus, 11¢ meant an increase of 13 3/4%. In Kuwait, producing 1,692,000 barrels a day, this amounted to $186,000 a day or $68 million a year.

Unlike many motorists, oil companies have always known that *fractions* of pennies add up to a tidy sum when selling millions of gallons of oil or gasoline every day. The next time you pull up to the pump to fill up your car, take a good look at the *actual* price. After the big bold numbers — there is always a teeny-weeny "9/10" of a cent. Americans guzzle over 350 million gallons of gas a day — *you figure it out.*

The Shah almost flushed fledgling OPEC down the commode. The companies made an offer to Iran contingent upon four members acceptance. Iraq threatened to withdraw from OPEC if the agreement was accepted in contravention of OPEC's rule that decisions should be unanimous. Iraq was supported by Venezuela and Indonesia even though the issue didn't apply to their concessions. To save face, OPEC deleted the vote from the agenda and allowed the members to handle the issue on their own, then passed a flowery resolution as to their solidarity and continuing struggle to obtain pre-August 1960 prices.

Many petroleum historians have written that OPEC accomplished little during the first seven years of existence. The more astute writers recognize that just getting organized and continuing to talk was a hell of an accomplishment...and it was also a learning experience.

Widespread dissatisfaction with OPEC spread in the Mid East, especially in nations without an abundance of oil. At the Fifth Arab Petroleum Congress in 1967, the out of work Red Sheik, Ṭariki, screamed for the nationalization of the oil companies and that Arab oil should be used as a weapon against those who did not support Arab causes, which didn't take a genius to figure out that it included anyone who supported Israel.

OPEC PARLIAMENTARY PROCEDURE

OPEC's rules provide, if the election of a Secretary General cannot be obtained by unanimous decision, the selection shall be on a rotation basis every two years. Unable to agree on a candidate in 1983, Iran insisted that an Iranian be appointed as every member had named a Secretary General and it was their turn again. Algeria objected, claiming that the rotation should now be by alphabetical order. (The Algerian delegate would not have objected if he was from Venezuela.) The result was that OPEC didn't get a full-fledged Secretary General until 1988, when they threw up there hands and named an Indonesian, because they seldom offend anyone. So much for unanimous decisions in a body politic — they're like hung juries.

16

SEDITIOUS SIXTIES STORIES — and Six-Day Slaughter

SYRIA SECEDES FROM UNITED ARAB REPUBLIC in 1961. Army officers rebelled and put the Egyptian Nasser sent to run Syria on a plane back to Cairo, then announced they had suffered a bellyful of Nasser and the UAR and wanted to run things on their own. However, the Syrians couldn't agree who should be in charge. Between 1961 and 1967, Syria had seven governments, including two military juntas, a National Revolutionary Council, a Presidential Council and one man twice, with a name too long to remember who didn't last a year.

NASSER SENDS TROOPS TO YEMEN in 1962 to support a rebel uprising against the royalist government. As expected, the Russians supported the Marxist revolutionaries with money and arms and Saudi Arabia and Jordan backed the royalists. Today, the war is laughingly called "Egypt's Vietnam." In 1967 Nasser had to agree to a cease-fire to pull his troops out of the mountain guerilla war because things were heating up with Israel.

In case you are forgot where Yemen is — it's a little country south of Saudi Arabia formed in 1990 by a loose collection of troublemakers, including the Peoples Democratic Republic of Yemen and the Yemen Arab Republic. It came close to breaking up in 1994 in a North-South style civil war. It supported Iraq during the Gulf War, so the Saudis sent 850,000 Yemeni menial workers home. Other than being a pain in Saudi Arabia's backside and ranking twenty-third in oil reserves in the world, there's nothing much else you need to know about the malcontents.

QUIZ: How can you spot a Yemeni?

Answer: All Yemeni men carry a big curved knife in their belt called a *jambiya* and their teeth are stained green from chewing *qat* — a green narcotic leaf that tastes so bad the DEA claims it will never catch on in America.

HILL TRIBES REBEL AGAINST THE SULTAN OF OMAN in 1964. Inspired by the Marxists next door in Yemen, the rebels raised hell until 1970 when the Sultan was overthrown by his son who promised reforms. Although the son, Qabus Ibn Said, still rules as an absolute monarch, the Sultanate of Oman remains under British protection. The United States maintains a standby military base in the sultanate in case things get hairy in the Mid East. Oman also received part of the Arabs' inheritance from God — it ranks eighteenth in world oil reserves. You should know that Oman's people are friendly to Americans.

SHAH ANNOUNCES THE "WHITE REVOLUTION" in 1962 to modernize Iran. His attempted appeal to the people with land reforms and granting women and non-Muslims the right to vote was opposed by **Ayatollah Ruhollah Khomeini.** The Shiite *mullah,* was offended more by the land reform than women's rights, as the Shiite clergy controlled vast land holdings. Khomeini was imprisoned then exiled to Iraq. (If the Shah had it to do over again, he would have ordered the Iranian secret police, the SAVAK, to kill the old SOB.)

IRAQ'S KASSEM KILLED IN COUP in February 1963. Only the local Baghdad press reported that young Saddam

Hussein returned from hiding in Egypt and rejoined his Ba'ath Party conspirators whose motto was "Unity, Liberation & Socialism." The Ba'aths only lasted until November before being chucked out. Saddam was sent to prison, but they made the mistake of letting him out in 1966. The Ba'aths elected him deputy secretary in time for the 1968 coup, which would lead to Saddam's reign of murder, terror and the mobster-like government enslavement of the Iraqi people. For the lack of anything nice to say about him, historians compare Saddam to Hitler but, as despicable as Hitler was, he had more class than Saddam... Well, at least Hitler's soldiers had prettier uniforms.

The Big Headlines in America – The Six-Day War

August 1963: Israel begins project to divert 75% of the waters of the Jordan River for agriculture and industrial development.

January 1964: Thirteen Arab nations meet in Cairo to determine what action to take to stop the water diversion they call "thievery." Nasser proposes an Egyptian general to lead a unified Arab command to protect Arab borders. As usual, it was regarded as Arab rhetoric — all talk, no action.

November 4, 1966: Egypt and Syria sign defense pact.

November 13, 1966: Israelis raid into Jordan in retaliation for the killing of three Israeli soldiers. (Israelis kill 18 Jordanian soldiers.)

December 1966 to April 1967: Border skirmishes between Israel and Syria, but Syria gets no help from Egypt. Nasser says: "What can I do?...The fighting is on the other side of Israel."

May 18, 1967: Arab bitching forces Nasser to insist that the United Nations Emergency Force, stationed in the Sinai since 1956 to keep the peace and insure free passage in the straits of Tiran, be removed. *Editorial Comment: If peacekeeping troops are forced to leave, isn't that a sign there will be no peace?*

May 22, 1967: Nasser moves the Egyptian army into the Sinai and shuts off the Gulf of Aqaba and Straits of Tiran to Israeli shipping.

May 30, 1967: Egypt and Jordan sign defense pact.

June 5, 1967: ISRAEL DESTROYS THE EGYPTIAN AIR FORCE ON THE GROUND IN PRE-DAWN RAID — THE START OF THE SIX DAY WAR.

June 6...Israeli army sweeps across the Sinai and reaches the Suez Canal on the 9th. Over 10,000 Egyptian soldiers killed or die of thirst.

June 7...Israel captures West Bank and Arab Jerusalem... Jordan agrees to cease-fire.

June 8...Israel attacks *USS Liberty,* an American electronic listening ship in international waters, killing 34 American sailors and wounding 171. Israel claims it was an accident, but no one believes them, especially after the second and third bombings. (Few Americans read that Israel had captured the Egypt's Gaza Strip.)

June 9...Egypt accepts cease-fire as Israelis storm the Golan Heights in Syria.

June 10...Syria accepts cease-fire.

The Six-Day War was a "turkey shoot."... There was no question who won the Third Arab-Israeli War. Nasser got his butt whipped and was humiliated. The Syrian and Jordanian armies demonstrated they should have stayed on the ceremonial parade ground, but that didn't mean there was peace...Terrorism increased. The real losers were not Egypt, Syria and Jordan, but 300,000 more displaced Palestinian refugees.

What Was OPEC Doing During the Six-Day War?

Nothing officially. On June 6 Saudi Arabia, Kuwait and Algeria embargoed oil shipments to "all countries helping Israel," namely the United States, Britain and, for a brief time, Germany. Libya and Iraq banned all exports in support of the Arab cause. To make sure there was no sleight of hand by Aramco, Oil Minister Yamani told the companies they would be in deep camel dung if they shipped one drop of oil to the United States or Britain. Strikes and sabotage temporarily halted oil production in Saudi Arabia and Kuwait; and mobs attacked American oil officials in Libya, requiring the U.S. Air Force to airlift them to safety.

The embargo of the affected nations amounted to 1.5 million barrels a day. Adding to the shortfall was a civil war in Nigeria, which blockaded another 500,000 barrels a day from the world market. However, it was only a matter of logistics. Supertankers came to the rescue when the Suez Canal was closed again and the pipelines from Iraq and Saudi Arabia to the Mediterranean were shut down.

But it wasn't an OPEC embargo: Iran, Indonesia and Venezuela turned loose an extra 700,000 barrels a day. In the United States, market demand prorationing was cast aside when the Texans and Okies uncorked an extra one million barrels a day.

In August 1967 the Arab leaders met in Khartoum, Sudan, to discuss damage control. To assure he would still be in charge when he got back to Cairo, Nasser ordered 150 army officers arrested to prevent a coup. Iraq demanded an embargo of the entire world to show the Arabs' oil power. Algeria and Syria refused to show up, aware their Arab brethren would cave in to the lure of the almighty petrodollars.[60] They were right. The embargo was lifted and threats of pulling all Arab money out of American and British banks simmered down. Big Oil was aware it was a foregone conclusion — Arab production in August was already running seven percent higher than it was *before* the embargo. Money did the talking. Saudi Arabia, Kuwait and Libya commended the "front line" countries of Egypt, Syria and Jordan and promised them subsidies.

---

[60] Syria attacked the sheikdoms as selfish, claiming Mid East oil was the "rightful property of the Arabs as a whole." In other words, the Syrian socialists wanted their share of Saudi and Kuwaiti oil.

The Six-Day War has been said to have caused the "third oil crisis," but it didn't. It happened so fast prices didn't get a chance to skyrocket. *The selective embargo didn't work because there was an abundance of crude oil in the world available to meet the shortfall.*

THE SHAH'S OIL DEAL

After the Six-Day War, the Shah visited Washington to remind Americans of his cooperation by pumping oil for his friends in their time of need. He had the bright idea that Iran should be given a special quota of a million barrels a day at $1.00 a barrel (half the going price) for storage in abandoned salt mines in case of another embargo. Everyone snickered. Within months of the embargo, there was a glut of oil on the market and predictions oil prices might fall. It would take another war and embargo for America to begin its Strategic Petroleum Reserve in abandoned salt mines of Louisiana and Texas.

As it turned out, the Shah had a great idea, and the price was right.

United Nations Resolution 242 — It Only Made Matters Worse

UN resolution 242, passed on November 22, 1967, had the lofty purpose of providing "a just and lasting peace... within secure and recognized boundaries." This brings to mind the old adage: *A good lawyer can find a loophole in the Ten Commandments.* However, any first year law student could have spotted the loopholes — there were no secure or recognized boundaries. Further, there was a phrase requiring Israel to withdraw its "armed forces from territories occupied in the *recent* conflict." What conflict did they mean?...this was the Third Arab-Israeli War! The Arabs and the USSR argued it should read simply "all territories" so there would be no questions, but were voted down.

Another aberration was the acknowledgement of the "sovereignty, territorial integrity and political independence of every state in the area," which sounded good to the Israelis — it was a de facto recognition of the state of Israel. From the Arabs' point of view, they didn't want to admit that Israel existed, sometimes called the "ostrich with its head in the sand viewpoint." *But the Arabs did have one good point: where was the Arab state? There wasn't anything left! Israel had grabbed pieces of Syria, Jordan, Egypt and the rest of Jerusalem.*

The compromise of the stalemated United Nations during the Cold War was an invitation for Nasser's next effort — the "War of Attrition," likened to two boys who start throwing rocks at each other...soon the rocks turn to bricks and one of them gets a gun...then the other gets a bigger gun...then the other gets his big brother...Soon both families are shooting at each other like the Hatfields and the McCoys. The Egyptians and Israelis sniped at the other across the Suez with rifles, then artillery, until the Egyptians saw American Phantom jets with Israeli markings. What was Nasser to do but go to Moscow and get a few new MiGs? By 1970 there were Soviet pilots in the MiGs. The only good thing you can say about the War of Attrition is that the Israeli pilots were good, so the United States wasn't conned into sending American pilots.

QUIZ: Did the Arabs think they got a fair shake by the United Nations and United States?

Answer: Hell no! The Arabs believed Israel was the aggressor — Israel attacked Egypt, Syria and Jordan and gobbled up all of Palestine, chunks of Egypt and Syria plus the West Bank...Israel's claim that it was a "pre-emptive strike" took a lot of chutzpah. The Arabs claimed they were "merely protecting their borders."

The Arabs said they never had a chance in the UN — the American ambassador to the UN, Arthur Goldberg...was a Jew...who failed to demand Israel's pull back to the pre-war borders, after delaying the UN cease-fire vote, so Israel could steal more Arab land.[61]

---

[61] Read *The Arab-Israeli Wars* by Chaim Herzog for the Israeli viewpoint and *Arabs & Israel for Beginners* by Ron David for the Arab side of the story.

17

OPEC & ARABS GET THEIR ACTS
TOGETHER — America Runs In Circles

As OPEC headed into the 1970s, it added three new members: Abu Dhabi (1967), which transferred its membership to the UAE in 1974, Algeria (1969) and Nigeria (1971), bringing its total to eleven.

The Arabs realized that the non-Arab members would not use oil as a weapon against the western nations, as shown when Iran, Indonesia and Venezuela screwed up their 1967 embargo by helping make up the shortage. After the January 1968 OPEC meeting, Saudi Arabia, Kuwait and Libya formed the Organization of Arab Petroleum Exporting Countries (OAPEC). By 1972 all the Arab oil producing countries were members. This didn't mean the Arabs were one big happy family. Syria, the constant whiner, shut down the pipelines within its borders from Iraq and Saudi Arabia to the Mediterranean and demanded an increase of three times what they were getting as a fee for the oil crossing Syria, but settled for double, which is par in negotiating — ask double what you expect to get.

A 1966 United Nations resolution gave OPEC a big boost when it called on all nations to acquire sovereignty over their natural resources, a diplomatic way of saying the hell with the sanctity of the contracts with Big Oil who had been screwing underdeveloped nations. Quoting the UN language, the June 1968 OPEC meeting adopted a resolution demanding "participation" in the ownership of the oil concession. Yamani told Aramco that 20% would be nice a start for Saudi Arabia, but rumblings were heard that a few nations wanted 100% participation, which was the Third World's way of saying "nationalization." The drawback of not knowing whether they could operate the oil fields wasn't important, there would always be plenty of Western technicians to keep the oil flowing and young Arabs have been going to the University of Texas and other citadels of petroleum technology for over a decade.

A Couple of Wild and Crazy Guys

The Arabs were also looking across the Persian Gulf (Arabs call it the Arabian Gulf) at Iran after Britain's announcement it was pulling its troops out of the Gulf. The seven small sheikdoms making up the Trucial States formed the United Arab Emirates in 1971 so the wouldn't be picked off one-by-one. The sheiks offered to pay Britain to keep troops in the area, but the British declined because they didn't want to admit they were broke. The UAE was worried about the Shah, who claimed Bahrain and several islands run by UAE sheikdoms no one wanted until oil was found in their offshore waters. The Shah backed off on his claim of Bahrain, but kept most islands.

Everyone knew the Shah had a big head. In 1971 he celebrated the 2,500th anniversary of the Persian Empire by crowning himself *Shahanshah* ("king of kings") at a gala that cost $250 million and was catered by Maxim's of Paris. By this time, America should have been concerned about the megalomaniac they were making the policeman of the Persian Gulf after the British pulled out. A year earlier the Shah had been in Washington asking President Nixon to con the American oil companies into lifting $150 million more in oil for his development programs. The Americans couldn't take any more oil and Nixon couldn't force them, which made the Shah more inclined not to permit democracy or free enterprise in Iran. The American companies only held 40% of the Iran consortium (BP, Shell and CFP held 60%) and were more interested in keeping Saudi Arabia happy.

If there was any question whether the Mid East was stable, it was answered in September 1969 when Muammar al-Qaddafi took over Libya in a coup. Qaddafi started off doing what revolutionaries are supposed to do — he organized a Revolutionary Command Council. Then he closed the American air base, tossed out any Italians still hanging around from the time they ran the desert wasteland and boarded up all the Catholic churches after selling off the crucifixes and stained glass windows as souvenirs.

For petroleum advisor, Qaddafi hired Abdullah Tariki, the Red Sheik, who had been canned from his last job in Saudi Arabia. However, the actual price negotiations were handled by twenty-seven year old Abdel Salaam Ahmed Jalloud, who paraded around with a machine gun slung over his shoulder and laid a forty-five automatic on his desk during meetings with the oil companies. If he didn't like an oil company executive's proposal, he rolled it up in a ball and threw it in his face.

Jalloud was hot to get a price increase based on his conception that Libyan crude oil was worth more than Persian Gulf oil because of its low gravity (true), sweetness (true), and was closer to Europe than the Gulf, which saves tanker costs (true). The fact that the Suez Canal was closed and a bulldozer "accidentally" cut the pipeline through Syria got his point across. However, he had an inflated idea that the increase should be between 35¢ and $1.00 a barrel, which Exxon said was ridiculous and countered with a similarly ridiculous 5¢ increase. Exxon could afford to stonewall Jalloud because it had more oil that it could market in Saudi Arabia, Iraq and Iran as did the other Sisters. But many independents didn't have an oil reserve surplus. Jalloud singled out Occidental Petroleum (Oxy) to put on the squeeze. Oxy's chairman, Armand Hammer, was told that until he agreed to a price increase of 40¢, Oxy had to cut back production from 800,000 to 500,000 barrels a day, which was followed by further cuts in order to wring Oxy dry.

When Hammer visited Exxon, begging for a cut-rate price in order to stave off Libya's price demands and meet Oxy's supply needs in Europe, it took a lot of chutzpah. Oxy was undercutting the prices of Exxon and the other Sisters in Europe with Libyan oil. Worse, a few years earlier, Oxy had committed an unpardonable sin. When Exxon's oil fields were expropriated by Peru, Hammer hopped on the next plane to make a deal to take over Exxon's Peruvian oil properties. Exxon told Hammer to get lost.

Hammer was a schlemiel and Exxon was pigheaded. Both would suffer. Oxy caved in to Jalloud's demands. The other independents had to join or **be nationalized.** Exxon and the other Sisters had to agree or else **they** would leave all **the** sweet, light Libyan crude oil a short distance from Europe to the little guys. The price was increased 30¢ a barrel plus 2¢ each year for five years.

Jalloud had another bitch — he claimed Oxy had been screwing Libya for five years under senile King Idris. To make up for past under-payments, the tax was increased from 50% to 55%. Jalloud and Hammer parted pals, committed to honor the agreement for five years. *As they say in New York, if you believe that, I'll sell you the Brooklyn Bridge.*

When the Shah heard that a backward son of an Arab camel herder negotiated a better price than an Aryan king, he blew his stack and demanded a price boost and increased Iran's taxes from 50% to 55%. Like dominoes, the sacred fifty-fifty agreements fell and prices increased under the Oxy formula through the Mid East.

Not to be outdone, Venezuela passed a law increasing its tax rate to 60%. The wily quiet Indonesians shrugged — they had been receiving 60% since 1963 but, as most of its sales were to the Japanese and Americans, no one had noticed. Both were rich countries that could afford to pay more for oil than the rest of the world.

ARMAND HAMMER: Portrait of an Outsider

Dr. Armand Hammer was never regarded as an oilman by Big Oil or "Little Oil." Most people held the opinion that he was an unscrupulous, money-grubbing wheeler-dealer. A few thought he was a closet communist. His father, Dr. Julius Hammer, was a founder of the American Communist Party. After graduating from medical school, Armand Hammer went to Russia with medical supplies and the idea he could schmooze the Soviets out of the $150,000 he claimed they owed his father for confiscating his pharmaceutical business. He became a friend of Lenin and operated several businesses in Russia, including the national concession for pencils. When

Stalin's anti-Jewish policy reared its ugly head, he pulled up stakes and left Russia.

Bored after retiring at age fifty-eight, he took over the almost bankrupt Oxy and parlayed it into a large integrated oil and chemical company. After striking oil in Libya, Oxy became the sixth-largest oil producing company in the world.

Armand Hammer was brilliant, ruthless and egomaniacal. He tried to take over Arm & Hammer, the well-known baking soda company, because of the similarity with his name, but only managed to acquire 5% of the then closely-held company. His other aim in life was to win the Nobel Peace Prize by making himself the unofficial goodwill ambassador between Soviet Russia and the United States and buying humanitarian awards with large Oxy corporate contributions. However, he only succeeded in making himself a fawning pain in the ass to five Presidents of the United States and four Presidents of the Soviet Union.

Armand Hammer Quotes

"I've always been famous because I knew Lenin...I want to be immortal...The only reason I give is to get."

"Those who insist on telling the truth never have a future. The only way to build the future is to build it on lies."

When asked about the environment, he replied: "Personally, I don't give a damn about it." Apparently, he told the truth for a change. One of Oxy's subsidiaries was the Hooker Chemical Corporation, which while it didn't make Armand Hammer immortal, it made the environmental disaster caused by toxic waste dumping at Love Canal go down in history.[62]

---

[62] For an uncomplimentary biography of Armand Hammer, read *The Dark Side of Power: The Real Armand Hammer* by Carl Blumay with Henry Edwards.

1970 News Dispatches From the Mid East

1970's biggest headline was a peace treaty between Egypt and Israel, which settled little, but at least stopped the War of Attrition.

King Hussein drove the PLO forces out of Jordan before the Israelis did any more damage to his little country by bombing Yasser Arafat's terrorists hiding in Jordan. The PLO ran off to Lebanon and established bases for their raids into Palestine and Israel.[63]

Hafez al-Assad took over Syria in a coup. Most people yawned at the news. It was the ninth coup in a decade, but Assad was still running Syria with an iron hand when 1998 rolled around.

Nasser died after suffering a heart attack and **Anwar Sadat** took over as President of **Egypt.** Nasser's funeral went into *The Guinness Book of World Records* as having the most mourners — 4 million.

Leapfrog

The OPEC meeting in December 1970 was a celebration of their victory in obtaining 55% as the minimum tax rate and an agreement to eliminate disparities in the posted prices. In other words, it was back to the bargaining table. This time, OPEC planned to divide and bargain as two entities, the Gulf and the Mediterranean producers.

---

[63] King Hussein's victory was a major disaster for the Palestinians, they call "Black September," but legend says it had its light moment. When the cautious King saw a bra flying from the radio antenna of a tank, he asked what it was. The troops are reported to have told him that they had been ordered to fight like women. King Hussein ordered them to charge and the Jordanians crushed the Palestinian guerrillas...The stuff Hollywood loves.

The American oil companies geared up for the battle by obtaining authority to act as a bloc from the Department of Justice to avoid antitrust violations and securing the support of the Department of State. The companies, aware the Shah and Qaddafi would attempt to outdo the other, causing the prices to "leapfrog" upward, attempted to maneuver for a single negotiation covering all Mid East suppliers.   State Department Under Secretary John N. Irwin sent to discuss the issue with the Shah, blew the plan because he received no instructions from Nixon. Like most State Department officials, he didn't know a hell of a lot about economics and even less about the petroleum business. All he was interested in accomplishing was not offending the Shah and assuring an uninterrupted supply of oil at reasonable prices, without any idea what was reasonable. As a result, he bowed to the Shah who insisted on separate negotiations in Tehran for the Gulf states and in Tripoli for the Mediterranean producers.

The Gulf producers increased their price 35¢ a barrel at a bargaining session in Tehran. Before venturing to the second round in Tripoli, the companies secretly agreed to create the "Libyan Safety Net;" providing, if Qaddafi cut a company's production or threatened nationalization, the others would make up any shortfall at their production cost. As predicted, the Tripoli negotiations led by Libya resulted in a price increase of 90¢ a barrel, jumping the price from $2.55 to $3.45 a barrel for the Mediterranean producers — 55¢ higher than the Gulf producers.

Separately, Algeria nationalized the French interests and raised the price to $3.60, Iraq nationalized the Iraq Petroleum Company, then Syria nationalized the pipeline within its borders leading to the Mediterranean. By 1972 "participation" was the main topic of conversation at OPEC meetings, with the stated objective to start with 25% ownership until it reached 51% in 1982, although some wanted 100% participation, which meant that some oil companies weren't going to "participate."

The Seven Sisters, now accompanied by over fifty oil companies of various sizes in the Mid East and Africa, no longer controlled prices and manipulated production to suit their needs. It didn't take an economist to figure out who the losers were. The price increases were passed on to the consumer, including an extra 8.49% in 1972 when the United States went off the $35.00 gold standard. OPEC raised the price for the devalued United States dollar, which Henry Kissinger thought "disingenuous."

In The United States, Things Turned Into a Fiasco

America was mired in the unpopular Vietnam, managed from the White House instead of the Pentagon for political reasons. President Lyndon Johnson saw the handwriting on the wall — he wouldn't be reelected, so he turned tail and ran home to Texas. America's other worries were rampant inflation and the environment.

When oil prices increased, the whipping boy was Big Oil. Big Oil was attacked for its exorbitant profits and favorable tax treatment — a foreign tax credit that allowed the Arabs to live high at American taxpayers' expense, a depletion allowance on oil lands they leased, not owned, and fast tax write-offs of drilling costs — that, when combined, permitted the oil companies to pay about one-fifth what other corporate taxpayers were paying the IRS. After Congress cut many of the tax deductions, Big Oil felt its profits being squeezed and raised prices at the pump, adding to the inflation.

Congress passed the National Environmental Policy Act (NEPA) in 1969. The Department of the Interior regarded NEPA as merely requiring loads of bureaucratic paper shuffling and studies that would never amount to much. Many members of Congress who voted for it felt the same way. Few realized NEPA's implications and its avenues for litigation to bar or delay energy programs.

In December 1967 oil was discovered in Prudoe Bay, Alaska — the biggest oil strike in the history of the United States, with promises of two million barrels of oil a day. The good oil news of the Alaskan bonanza was overshadowed by a blowout on an oil platform off the coast of Santa Barbara, California, in January 1969 that spilled 70,000 barrels of oil. Pictures of pelicans trapped in goo and black slimy beaches alarmed the nation and caused drilling in the offshore area to be curtailed.

Despite warnings from Interior Department officials that the nation was becoming dependent on oil from unstable Mid East nations, the opening the giant Alaskan field was delayed time and time again for environmental and political reasons. The Johnson administration placed a freeze on the construction of the Trans-Alaska Pipeline System (TAPS) until the federal land claims of the Alaska Natives (Eskimos, Indians and Aleuts) and the State of Alaska were settled, which would drag on for three years.

ALASKAN GIVEAWAY

Approximately 40,000 Alaska natives clamored for a grant of 40 million acres of Federal land (about the size of Wisconsin) and $1 billion in cash.

The State of Alaska was given 102 million acres (about the size of California) as a reward for becoming a state plus 90% of the oil royalties from all Federal lands (compared to 50% received by all other states). Every rugged individualist citizen of Alaska received a $1,000 check in 1982 and continues to receive periodic handouts as a result of the boondoggle. The state has no income tax because of the generosity of the great federal giveaway.

TAPS' environmental safeguards and alternatives were discussed ad nauseam, including a hair-brained scheme to transport crude oil under the ice by submarines. Vitriolic politics by the ranking Republican on the House Interior Committee, John P. Saylor of Pennsylvania, turned the committee into a circus when he learned that TAPS had purchased Japanese pipe for the 789 mile pipeline and had not bothered to get bids from American steel companies, especially those in Pennsylvania that were suffering from stiff competition from the Japanese.[64] Saylor delayed the pipeline proceedings almost two years. The Interior

---

[64] Representative Saylor was the former president of Ringling Brothers and Barnum and Bailey Circus. For an engrossing muckraking sideshow of the Alaska pipeline delay and America's first oil shortage, I suggest *Fiasco*, by Jack Anderson, with James Boyd. The title personifies what was going on in Washington.

Department further delayed progress when it issued a wider pipeline right-of-way than permitted under the law and was enjoined in court by environmentalists.

TAPS would not begin construction until February 1975, and the first oil would not flow through the pipeline until June 1977, over ten years after oil was discovered and too late to help during America's first real oil shortage and effective oil embargo by the OAPEC.

QUIZ: Which of the Seven Sisters is the largest producer of federal oil in Alaska?

Answer: Believe it or not: British Petroleum.

When Richard Nixon was sworn in as President in 1969, crude oil prices in the United States were $1.10 a barrel higher than the world market price. One of Nixon's first actions was appoint a Cabinet task force to study the quota system; however, the Cabinet couldn't agree. Although the majority recommended scrapping the quotas for a tariff, a strong minority led by the former governor of Alaska, Interior Secretary Walter J. Hickel, objected on the grounds that it would drive the price of domestic oil down from $2.90 to $2.40 a barrel and cripple the American oil industry. The minority, supported by heavy oil industry lobbying, won.

At the same time, the nation was facing a related issue — so-called brownouts — shortages of electricity caused by the lack of power that would blackout the New York metropolitan area, leaving it without electricity for several days. By the mid-1970s many schools and industries in the East and Midwest would be forced to close during the winter because of a shortage of politically inspired government-controlled "interstate" cheap natural gas, while the utilities and industries of the oil and gas producing states using uncontrolled "intrastate" gas had plenty, but were paying about $1.00 Mcf more. The political and bureaucratic ineptitude that had been going on for decades caused a natural gas short-age in a nation that had an abundance of the clean burning fuel.[65]

In 1970 American oil production reached its peak at 11.3 million barrels a day, which has declined to 6.3 million barrels a day in 1998. Import quotas contributed to maintaining domestic prices 60% higher than the world market. Market demand prorationing was no longer nec-essary to maintain domestic prices. By March 1971 the Texas Railroad Commission was only restricting production for conservation purposes. The cushion that saved the United States from the Arab embargo in 1967 had disappeared. Today, there are few places left to search for new oil fields in the United States except in the Outer Continental Shelf and Alaska, both of which are hampered by environmental constraints.[66]

What Did Washington Do About the Mess?

Play politics is the short answer. In response to the inflation, Congress passed the Economic Stabilization Act of 1970 (ESA), autho-rizing the President to "stabilize prices, rents, wages and salaries." Congressional Democrats hoped that Republican Nixon, a strong oppo-nent of price controls, wouldn't use the legislation to combat rising costs while publicly screaming that he should. Nixon phased out the import quota system and replaced it with import fees (an underhanded term for tariff), but it wasn't enough. To everyone's surprise, Tricky Dick imposed price controls and established the Cost of Living Council to combat inflation in August 1973. Reacting to oil shortages in early

---

[65] A review of the natural gas dilemma is beyond the scope of this primer. For those interested in an insightful study of the issues, I suggest *The Natural Gas Industry: Evolution, Structure and Economics* by Arlon R. Tussing and Connie C. Barlow.

[66] In 1995 California chose the environment over oil by banning oil production in the state's coastal waters. The Californians' love of the automobile has made Los Angeles the nation's smoggiest city, which also means they chose their cars over their lungs.

1973, particularly in the politically powerful Northeast and Midwest where independent oil refiners and distributors were without sufficient supplies of oil, Congress amended the ESA:

> "...to promote and maintain competition in the petroleum industry and assure sufficient supplies of petroleum products to meet the essential needs of various sections of the Nation, it is necessary to provide for the rational and equitable distribution of those products..." and "set priorities of use and systematic allocation of supplies."

The high-minded language meant that Texans, Okies, Louisiana Cajuns and other oil producing states had to share their oil and gas with the damnyankees and, in effect, subsidize oil and gas costs under a government boondoggle. The Cost of Living Council regulations were too complex, too political and too pea-brained.

QUIZ: Did the price controls work?

Answer: As any first-year economics student knows, price controls *never work*. They destroyed competition by encouraging inefficient domestic oil production and allocating supplies while stimulating consumption with artificially low prices, which compounded the shortage and added to inflation.

At the White House, Nixon was embroiled in the Watergate scandal that would eventually drive him from office. He was also trying to avoid admitting he even knew Vice President Spiro Agnew, who would be forced to resign in October 1973 for alleged payoffs received when he was Governor of Maryland.

Secretary of State Henry Kissinger later admitted he didn't understand the significance of oil in the Mid East in 1973, other than he didn't want the Russians to get their hands on it, although he was pushing the catchword "détente". In 1971 Kissinger acknowledged he followed the time-honored government practice of ordering a State Department study of oil's national security implications, which no one confessed to writing or reading. Until 1973 "energy was considered a domestic, not a foreign issue," according to Kissinger.[67]

The above helps explain why the United States government stood by and permitted the leapfrogging of prices in the Mid East[68] *and why America was unprepared for what was to occur.*

---

[67] Kissinger's *Years of Upheaval* is an excellent chronicle of American foreign policy and a must reading for those desiring to understand our Mid East policy during the 1970s. Kissinger should be credited for being one of the few men in politics to admit his shortcomings and mistakes in his memoirs: "We knew everything but understood too little. And for that the highest officials — *including me* — must assume responsibility."

[68] The State Department official who told the Shah and King Faisal that the United States government wouldn't intervene in the oil companies' negotiations in the Tehran and Tripoli leapfrogging gave Iran and the Arabs the green light to raise prices and squeeze the American consumer and destroy the economies of the Third World nations.

18

THE FOURTH ARAB-ISRAELI WAR —
The Oil Weapon

Who is This Guy Sadat? And What is He Up To?

At first, Egypt's new President, Anwar Sadat, quietly followed Nasser's policies. Upon realizing no one was taking him seriously in his desire to rebuild the nation and uplift its image, he fired his vice president and, when seven ministers resigned in protest, he had them arrested and charged with an attempted coup. His show of strength and open policies increased his popularity, but only marginally. Egyptians had not gotten over their disastrous defeat by Israel in 1967.

Israel's occupation of the Sinai deprived Egypt of desperately needed revenues from the Suez Canal and oil fields in the northern Sinai. America's lukewarm support in 1970 under the Rogers Plan (named after Nixon's first Secretary of State) required major Egyptian concessions, but offered Egypt little but smoke and cloudy mirrors. Sadat's 1971 offer to open the Suez Canal in return for Israel's partial withdrawal from the Sinai failed to get Nixon's support or a response from Israel — it loved the idea of pumping Egyptian oil from the captured Sinai

Sadat's only way to regain the Sinai was militarily, but his request to the Kremlin for weapons was rejected and Soviet loans were minimal. Sadat showed his anger and daring in April 1972 when he flew to Moscow and told Leonid Brezhnev that the United States supported Israel more than Russia supported the Arabs, and that Russia was hurting the Arabs by sending more Jews to Israel. Three months later, Sadat ordered the 15,000 Soviet military and civilian personnel to leave Egypt within one week.

The White House cheered the move and made token overtures to Sadat, but was too busy pressing détente with Russia to provide meaningful help. Also, Nixon wasn't about to arm Egypt with a strong Jewish lobby looking over his shoulder during an election year. Thus, Sadat was without a superpower to support his ambition to force Israel from Egyptian territory.

But Sadat was not alone. He had Hafez Assad of Syria to assist him militarily for a joint attack on Israel from both sides. Syria's Golan Heights was still occupied by Israel and, if anyone in the neighborhood hated Jews, it was Assad.

The Egyptian President pleased King Faisal of Saudi Arabia when he ejected the Russian communist menace from next door. Faisal also loathed Zionism and would bring money to the war — $500 million — and the biggest weapon of all — *oil.*

EGYPTIAN PROVERB (JOKE)

After the war, Egyptians loved to joke: "Saudi Arabia will fight Israel until the last Egyptian is standing."

Sadat's plan was simple, but far from foolproof: Egypt and Syria, joined by a few Arab friends, would fight a limited war to regain the borders stated in United Nations resolution 242. The Arabs would have the advantage of surprise and take the lands, then the United States and Russia would force a cease-fire based on the status quo, just as the Israelis had done in 1967.

Popular thought is that the infamous Yom Kippur War came as a total surprise on the most holy of Jewish holidays and during the Muslim fast of Ramadan...That's a big lump of camel dung! Admittedly in the days before spy satellites, the CIA and Israeli Mossad may have been half asleep when they failed to notice three Iraqi divisions moving into position in Syria and a Moroccan brigade that slipped into Egypt. (If you remember your eighth grade geography, Morocco is 3,000 miles from Syria in northwest Africa.) The United States and Israel were aware of Egyptian and Syrian troops moving towards Israel's borders.

On October 5, the Israeli military went on full alert (its highest degree of preparedness, with over 100,000 troops mobilized) after its military intelligence saw Egypt's troop build up. The same day, the CIA reported that the Russians had been evacuating their civilian personnel from Egypt and Syria for two days. The problem was that few understood Sadat or believed the puny Egyptian and Syrian armies planned to attack on *October 6, 1973.*

The Set Up...The Sting?

Between 1970 and 1973, United States oil imports rose from 3.2 to 6.2 million barrels a day — almost double. Worldwide oil consumption was increasing 7 1/2% a year. In the summer of 1973, economists and politicians in America debated whether there was an "energy crisis" or "oil shortage" based upon the basic law of supply and demand. Under this concept, "market price" means what a knowledgeable buyer and seller are willing to buy and sell a commodity when not under duress. By August panic buying by Americans, Europeans and the cash-rich Japanese pushed the spot market price over the Tehran agreement prices. Afterwards, economists debated whether the panicked buyers were knowledgeable, under duress or just plain stupid. So much for economic definitions.

The price hawks, Algeria, Iraq and Libya, demanded price increases at the June 1973 OPEC meeting. Saudi Arabia and other so-called moderates prevented an immediate price increase because of the soaring world inflation, but agreed to reevaluate the Tehran price agreement at the next meeting. At the September 1973 OPEC conference, the Tehran and Tripoli prices were declared dead and long overdue for burial. OPEC's evaluations also disclosed that the oil companies were reaping what they considered obscene windfall profits from Mid East oil. Yamani of Saudi Arabia called for renegotiations of the Tehran price agreement in Vienna on *October 8, 1973.*

On October 5 Big Oil finally obtained the necessary clearance from the Department of Justice to deal as a block, but it took some doing. The DOJ Antitrust Division, which saw an antitrust conspiracy behind every oil well, insisted the increased prices were the result of a rip-off by Big Oil and not due to OPEC demands or market conditions. In other words, the DOJ knew Big Oil was making humongous profits but didn't know how. (If the DOJ lawyers were that smart, they would have been hired by an oil company an triple their government salary.)

SURPRISE! SURPRISE! When the OPEC delegates and oil company representatives arrived in Vienna, they were told that the Fourth Arab-Israeli War had started and the Arabs were beating the hell out of Israel. The Arabs denied the conspiracy theory and historians avoid it, except to note that Sadat and King Faisal knew when the war and Tehran meeting were scheduled to begin. Thus, the timing of the war went down in history as a happy coincidence or horrible twist of fate, depending on your viewpoint.

Big Oil went to the meeting with a ceiling of a 60¢ per barrel increase set by their corporate headquarters, which would have raised the Gulf price to roughly $3.50. At the outset, they offered 45¢, a healthy 15% increase. The OPEC representatives gave them the finger and demanded a 100% increase. They wanted $6.00 a barrel!

The company negotiators called their headquarters in the United States and Britain for instructions, but all they heard on the other end of the line was: "Holy shit!" Big Oil expected to take some heat for any price increase, but doubling the price would bring the wrath of Ralph Nader and Ted Kennedy down on them; therefore, they asked for a two-day recess to consult with their governments. Uniformly, the United States, Japanese and European politicians screamed bloody murder, but offered no advice other than: "Hang tough, boys, we can't afford it."

In the wee hours of the morning of October 12, the weary Exxon and Shell representatives asked Yamani for two more weeks to conduct further consultations with their governments. Saudi Arabia's Yamani told them they were in deep goat shit and to listen to the radio. Then the Gulf OPEC representatives got on the next plane to Kuwait.

On October 16, the unified Gulf oil ministers from Saudi Arabia, Kuwait, Iran, Iraq, Qatar and Abu Dhabi announced the new Gulf price was $5.11 a barrel. The ministers took a page out of the oil companies' books by not rounding off the price to the penny, but added nine-tenths of a cent — the actual posted price was $5.119. At that price, who would notice the teeny-weeny *9/10¢?*"

The world was shocked at the 75% price increase. OPEC nations replied that they were merely adopting the market price and reminded the European governments that they were taxing gasoline made from OPEC oil at rates in excess of the OPEC price..."We're the ones that should be bitching!"

The oil companies were unnerved for another reason. Gone were the days when they could fix the price of Arab oil unilaterally as they saw fit; and they no longer could drag out the sticky negotiations with the Mid Eastern nations — *the OPEC Brothers could now set the price without so much as asking the Seven Sisters to the dance.*

OPEC BITCH — FACT OR FICTION?

OPEC's bitch was the European nations were taxing gasoline at a rate higher than OPEC's price, taxes and royalties, including the company's production, shipping, refining and marketing costs and *profits*. The complaint is still valid. Today, European nations and Japan tax gasoline more than the combined oil price and costs. America's love of the automobile turns Congress' knees to rubber when it votes to tax gasoline. Americans pay one of the lowest gasoline taxes in the world outside the OPEC nations, which subsidize the price of petroleum products to their citizens. *Figure 4* compares the costs of gasoline in the United States and several other industrialized nations.

In 1996 United States gasoline taxes included 18.3¢ per gallon federal and an average of 23.7¢ in state taxes. Some state and county taxes exceed 45¢. During the next election, write your governor and ask what you are paying in state gasoline taxes.

Figure 4

GASOLINE PUMP PRICES — MARCH 1996 (gal.)

	Cost	Tax	Total
United States	.88	.42	1.30
Canada	.84	.79	1.63
Germany	.89	2.97	3.86
France	.95	3.54	4.49
Britain	1.37	2.31	3.68
Japan	1.92	1.93	3.85

Organization for Economic Development and Cooperation

The Oil Gun

As the Iranian oil minister was leaving OPEC confab in Kuwait, the OAPEC delegates arrived for their meeting. The only item on the agenda was how to use oil as a weapon of war. The Iraqi oil minister, on orders from Saddam Hussein who had recently signed a security treaty with the Soviets, wanted to nationalize the American companies, pull every Arab dollar from American banks and halt oil shipments to America and other nations supporting Israel. When cooler heads prevailed, the Iraqi minister stormed out, warning that Iraq would not be bound by anything OAPEC decided, including an embargo. Back home in Iraq, Saddam was busy nationalizing the interests of Exxon, Mobil and Dutch-controlled Shell in the Iraq Petroleum Company.

The remaining OAPEC members, Algeria, Abu Dhabi, Bahrain, Egypt, Kuwait, Libya, Qatar, Saudi Arabia and Syria announced a progressive 5% cutback on oil production each month until an "evacuation of Israeli forces from all Arab territory occupied during the June 1967 war is completed and the legitimate rights of the Palestinian people are restored." A few OAPEC members liked the idea so much, they volunteered to cut production 10% each month.

Sadat's plan looked like it might work for a few days. When the Arabs' advances started to bog down on October 10, Russia began airlifting supplies to Syria and Egypt. The Israelis, unprepared for the Arabs' early victories, called on the United States for weapons. Détente went to hell. American policy was simple: we couldn't let the Russians back the winning side and the Israelis were our friends and swayed more votes than the Arabs at election time.

While OAPEC was meeting in Kuwait, Nixon and Kissinger were conferring with the Arab ambassadors in Washington, including Ambassador Omar Saqqaf of Saudi Arabia, who confirmed Israel's right to exist, but only within the 1967 borders. The President swore his support of UN Resolution 242 and told the ambassadors that the reason the United States was supporting Israel was because the real fight was between the United States and the Soviets.

Ambassador Saqqaf nodded that he understood. Whether he really did is not clear. The polite Saudi was too embarrassed by Nixon's mention that Kissinger would be America's negotiator with Israel to resolve the UN Resolution 242 issues...but not to worry that he's a nice Jewish boy. (Henry admitted that he was red-faced, too.)

After the meeting, Kissinger told everyone in the White House he was pleasantly surprised the Arabs never mentioned oil and that he doubted they would use oil as a weapon against America. That was his first mistake.[69] From that point, it is suspected the White House planners were Groucho, Harpo and Chico. The Israeli weapons air lift plan called for Israeli El Al Airlines planes, but after it was determined El Al didn't have the capacity, American commercial airlines were asked to do the job. The airlines said that there were no "friendly skies" in the Mid East and their pilots didn't like to be shot at. It was then decided to use U.S. Air Force C-5As to fly the supplies to Tel Aviv under the cover of darkness when no one would notice...really. It mattered little that there was a SNAFU in

---

[69] Kissinger wrote in *Years of Upheaval:* "At that time I had no experience with Saudi Arabia at all or the indirect, adaptable, and subtle method by which the Saudi policy was conducted." He learned a few weeks later that the Saudis are not always subtle when he threatened "counter measures" against the Arab embargo. Yamani replied that United States military action would be suicide because the Arabs would blow up all their oil fields.

obtaining permission from Portugal to refuel in the Azores, causing the planes to arrive in broad daylight and in plain view — on the same day, October 19, Nixon announced $2.2 billion in military aid to Israel.

The next day, the pissed off Arabs, led by Saudi Arabia and Libya, announced a total oil embargo against the United States and all countries supporting Israel, which would eventually include the Netherlands, South Africa, Rhodesia and Portugal, the latter for allowing the air lift to refuel in the Azores.

Kissinger learned of the Arab embargo while on his way to Moscow to negotiate a cease-fire. In unbelievable doublespeak, he rationalized: "No one believed that military urgency prevented us from exploring the method of resupplying Israel that would least jeopardize our interests in the Arab world and the dependence of the industrial world on imported oil." Almost in the same breath, he continued, "I wanted a demonstrative counter to the Soviet airlift." Henry was so busy worrying about the Soviets, he obviously missed King Faisal's rare interviews in *Newsweek* and the *Washington Post* and on NBC television a few months earlier when Faisal warned that America's continuing support of Israel against the Arabs would jeopardize the friendship between the nations and make it impossible to continue supplying America with oil.

Was the White House blind to the fact that the Arab people would have rioted in the streets and revolted if their governments failed to take action? Probably. Nixon had other things on his mind. The "Saturday Night Massacre" took place the same night after Nixon fired the Watergate special prosecutor for subpoenaing the secret Oval Office tapes and the Attorney General and his top deputy resigned in protest.

Kissinger negotiated a cease-fire in Moscow, which was accepted by both sides on October 22. However, everyone didn't get the word. The Egyptians and Israelis kept shooting at each other. When it appeared that the Egyptian Third Army would be annihilated, Russia threatened to intervene unless joint American and Soviet troops were sent to stop the slaughter of the Egyptians. The United States put its military on standby and went on a nuclear alert before Brezhnev backed down and the fighting finally stopped on October 24 or 25, depending on who you believe fired the last shot.

During the crisis Nixon was too shook up over Watergate to attend meetings or discuss the crisis with other world leaders, so he went to bed. The world would have to wait for Kissinger to complete his famous shuttle diplomacy between Israel and the Arab nations to determine if he could iron out UN resolution 338, which called for the implementation of resolution 242 "in all its parts," whatever that meant. *This was important...the Arabs had not lifted the embargo!*

And the Oil Gun is Pointed at Your Head

The embargo caused prices to skyrocket. The free world faced a real shortage of oil for the first time. In December Iran held an auction to determine the market price and was astounded to find it brought $17.04 a barrel — over triple the OPEC bench mark price. The following week, Nigeria held a rigged auction in which a desperate inexperienced Japanese firm was flimflammed into bidding $22.60 a barrel. That could only mean one thing — OPEC had to raise its prices.

The Gulf producers decided to meet again in Tehran to set new prices, but the other members decided to show up, too, because Saudi Arabia was mumbling about a price of $8.00 a barrel and its concern that any greater increase would cause a worldwide depression.

The Tehran meeting started off with OPEC's economists recommending $23.00 a barrel based on the inflated Nigerian scam and Saudi Arabia's counter of $8.00, which was plucked out of the air. However, it was America's friend, the Shah, who ran the show. The Shah determined he needed to net $7.00 a barrel for his national budget, so he worked the numbers backwards and recommended the posted price be set at $11.65.

The Shah explained that his computations were based on the cost of alternate forms of energy, such as coal and natural gas. This was the ideal increase to stifle coal and gas as competition with Mid East oil. It was pointed out that President Nixon had announced "Project Independence" to make America totally free from foreign oil through a myriad of research programs, conservation and legislation serious economists and oil industry officials laughed at.

Like most government programs, Project Independence consisted of political smoke and mirrors. It's only visible evidence were pseudo scientists' plans to save the world and congressmen clamoring for billions of tax dollars to be spent reinventing the wheel in their congressional districts. When Nixon heard that Kissinger was having a hard time getting the Arab embargo lifted, he ordered gas rationing stamps printed.

On December 23 OPEC adopted the Shah's suggestion of $11.65 to take effect January 1, 1974. *Between February 14, 1970 and January 1, 1974 — less than four years — oil prices increased 547%! In ten weeks, the price of oil had jumped 300%!*

OPEC IS TOO ARAB

At OPEC's November 1973 meeting, it admitted Latino Ecuador as a member and tiny black African Gabon as an associate member (just sit at the meetings and smile at the cameras, you're to small to vote) to remove the heavy Arab flavor from the organization. It was the equivalent of the New York Yankees signing two Little Leaguers. Ecuador and Gabon didn't go on the Tehran road trip, knowing they could raise prices to the same levels as the big hitters. The fact that Ecuador and Gabon are no longer OPEC members shouldn't surprise anyone.

Figure 5

OPEC POSTED OIL PRICES & GOVERNMENT REVENUES[70]

Effective Date	Saudi Light 34° Marker Crude Price	Average Government Revenue
1960 through		
13 November 1970	$ 1.80	0.91
14 November 1970	1.80	0.98
15 February 1971	2.18	1.27
20 January 1972	2.48	1.45
1 June 1973	2.90	1.82
16 October 1973	5.12	3.45
1 January 1974	11.65	9.31
Petroleum Economist		

America learned who its true friends were. So did Israel. As usual, it was "Uncle Sap" who paid through the nose. Yamani made it clear: *if you support the Arabs, you get Arab oil; if you don't, you don't get Arab oil.* The industrial nations' ineffectual answer to OPEC, the Organization for Economic Cooperation and Development (OECD), fell apart as each country made its peace with the Arabs. Only the stalwart Dutch and the United States hung tough. The OECD did manage to form the International Energy Agency (IEA) in 1974, made up of twenty-one nations, pledging their cooperation to reduce their dependence on

---

[70] OPEC pricing was beset with intricacies far beyond the scope of this text, such as buybacks, netbacks, participation percentage calculations and quality differentials. See Ian Skeet's *OPEC: Twenty-Five Years of Prices and Politics,* if you are interested in the nitty-gritty of international oil pricing, including the quirks in the OPEC 1973 price increases, which resulted in giving some companies a profit margin of up to $2.88 a barrel until OPEC woke to the fact and eliminated the oil companies' windfall.

oil by setting up storage reserves and the sharing of available oil in the event of an emergency. Translation: they did nothing. The United States, Germany and Japan were the only nations to actually establish government reserves. The spineless French refused to join the IEA for fear the Algerians and other Arabs would cut off their oil.

Like all international organizations, each IEA member had a different itch to scratch. Japan, totally dependent on imported oil, publicly embraced the Arab cause, having no choice but to bow low and say, "So sorry!" In varying degrees, the Europeans supported the Arabs and distanced themselves from the United States. Our "friends" pointed out that even though America swills one-third of the world's oil, its vast domestic production actually made it less dependent on imported oil than the European nations that had little or no oil; thus, America's average per barrel cost was much lower and gave America an economic advantage. *The bottom line is, when oil is involved, don't rely on old alliances and friends, especially the French.*

American military might didn't count, as Secretary of Defense James R. Schlesinger discovered when he blustered that the Arabs were risking violence. The answer from Kuwait was that they would blow up the oil fields if America used force.

Crafty Anwar Sadat became pivotal, bargaining for a favorable settlement for Egypt in return for lifting of the embargo against the United States. Overnight, Sadat became America's friend...*for a price.* The Syrians, who ended up losing territory during the war, were impossible to please. Israel offered the Arabs the bait: "A piece of territory for a piece of peace." King Faisal of Saudi Arabia listened stoically — he was in no hurry to take the United States off the black list. It was the Americans who were short of oil and in the position of having to satisfy Israel, Syria and Egypt. Nixon and Kissinger didn't dare tell Israel to retreat to the 1967 borders, it would look like a Soviet victory and might cost votes at home. Kissinger sent the author to the dictionary when he described the "powerful tergiversations of Saudi Arabia" in the lifting of the embargo.[71]

---

[71] According to Webster, Kissinger meant *powerful* evasions, equivocations or subterfuges — Kissinger had not heard the Arab proverb: *Trust in God, but be sure to tie your camel.*

Kissinger's diplomacy didn't succeed until May 1974 when Saudi Arabia needed Yankee dollars and Sadat was happy.

Kissinger probably deserved the Nobel Peace Prize for the peace settlement. (Hell, later they even gave it to Arafat.) Although he had been Nixon's National Security Advisor for four years, he wasn't confirmed as Secretary of State until two weeks before the outbreak of the war. However, his claim that "we demonstrated we could resist the vaunted oil weapon" is gasconade Pecksniffery. (My turn to send readers to the dictionary — I could have said: "a crock of shit.") The Arabs thought they won, at least they had the United Nations talking about the 1948 borders and they jumped from being rich to super rich.[72] Saddam Hussein, not bound by the OAPEC embargo, filled his coffers by embargoing only the United States and the Netherlands and turning on the oil spigot for every other oil-starved buyer. Sadat got something — he would open the Suez Canal and get the Sinai oil fields back. Increased prices were great for the Shah's pocketbook. Our Cold War foe, Russia, sold oil at inflated prices to a world begging for oil, including the United States, which fueled the Cold War for over another decade. Even Kissinger admitted the world was facing an economic crisis cased by the embargo and prices.

QUIZ: Did the OPEC's quadrupling of oil prices in ten weeks contribute to the worldwide inflation?

Answer: OPEC denied it. At a 1975 meeting of OPEC heads of state, OPEC reported: "The unilateral decision by OPEC Member Countries to adjust their oil prices had contributed but insignificantly to the high rates of inflation which has been generated within the economies of the developed countries." **OPEC lied.** Although the worldwide inflation was in full swing when they put the oil weapon to the heads of the indus-

---

[72] Assad believed Syria was the big loser and that Sadat brown-nosed the United States to get more favorable treatment for Egypt. He never forgave Sadat, nor would he forgive Israel for the wanton destruction of Quenitra, the provincial capital of Golan, before they withdrew.

trialized nations as well as the poor underdeveloped countries, OPEC gave inflation a tremendous goose.

Economists love to tell the public that rising oil prices by themselves do not cause inflation, but will admit they contribute to inflation. When economic progress slowed during the 1970s, economists came up with the term "stagflation." President Carter's top inflation fighter, Alfred Kahn, who wrote the universally acclaimed *The Economics of Regulation,* was not allowed to use the word inflation, so he called it "bananas." After complaints from the banana importers, he called it "kumquats."

However, no economist had the nerve to tell a cab driver in 1974, paying four times what he paid for gas two months ago and unable to raise his fares because of local price controls, that there wasn't inflation. The widow living on a fixed income, who didn't have enough money to pay her oil heating bill, knew the cause of the inflation was oil prices.

God created economists to make weather forecasters look brilliant.

QUIZ: Was the Yom Kippur War the Fourth Arab-Israeli War?

Answer: It depends on how you count.

1st — 1948 Palestine War

2nd — 1956 Suez War

3rd — 1967 Six-Day War

4th — 1969-70 War of Attrition was sort of a war

5th — 1973 Yom Kippur War/Ramadan War

6th — 1981-? Lebanon War that's really still going on. Hezbollah still shoots an occasional rocket at the Israelis and the Israelis still drop bombs in Lebanon. Israel has a neat name for the war: "Peace for Galilee."

19

AFTER THE OIL EMBARGO — Just When You Think Things Can't Get Any Worse, They Do

Events in the 1970s awakened America to the connection between the Mid East and oil. Long lines circling the block to fill up gas-guzzling cars and signs at the gas stations — "NO GAS TODAY" — pissed off Joe Sixpack, who earned his name by driving two blocks to the 7-Eleven to pick up a six-pack of Coors on cheap gasoline every evening. The price at the pump was spiraling, and it was easy for Joe Sixpack to blame Big Oil and the Arabs after the 1973 embargo. Our nation's economic and political experts (who my mother called "know-it-alls"), had steered him down complacency highway with their foot on the accelerator.

In 1968 the Brookings Institution, one of the nation's lofty think tanks, prepared an "analysis of the substantive problems with which the new President and new Congress would have to grapple" under a grant from the Ford Foundation. Oil was merely mentioned in passing by one of the eighteen know-it-alls in their high-toned *Agenda for the Nation*. John C. Campbell, Harvard Ph.D., a senior research fellow at the Council on Foreign Relations, vice president of the Middle East Institute and former State Department staffer (all academic credentials, and not one drop of sweat worrying about meeting a payroll or producing a barrel of oil), pontificated:

Because oil is often cited as the one reason for America's presence in the Middle East, a few words may help to put it in perspective. The Middle East, with Arab North Africa, accounts for over 40% of the noncommunist world's oil production (1967) and contains about two-thirds of the entire world's proven reserves. Produced in large part by U.S. companies, it is important in terms of investment, revenue, and the balance of payments. *But it is not essential as fuel to the United States. (Emphasis added.)*[73]

Dr. Campbell and the Brooking Institution were not alone in their blundering. Only the Pentagon charged with fueling the next war and a few voices in the tree-hugging wilderness of the Interior Department warned of the hazards of foreign oil dependence.[74] (The next time a Ph.D. or media pundit pontificates, read what they said years earlier before you believe them. Most would do better with a crystal ball.) The so-called experts also mumbled the trite phrase that the Mid East "was a tinder box," which meant that it could go "boom" any minute.

Americans had heard murmurs of the horrors of war and terrorism in the Mid East, but it wasn't until one evening in September 1972, as they sat on the sofa sipping a cold beer in the comfort of their homes, did Joe Sixpack witness it on television. Suddenly, live from the Olympics in Munich, Germany, eleven young Israeli athletes were gunned down by the Black September, a Palestinian terrorist organization. Few knew Black September was organized to avenge the defeat of the PLO in Jordan by King Hussein — an Arab. It didn't make any sense to the Sixpack family, nor did the later air raids by the Israelis on refugee camps in Lebanon or the airline hijackings and taking of hostages by the Palestinians. The tinder box metaphor was passé. Much of the area was already in flames and ashes.

Joe Sixpack shook his head at Walter Cronkite's analysis of the realpolitik of the Lebanese civil war and the plight of the Palestinian

---

[73] Brookings Institution. 1968. *Agenda for the Nation.*

[74] Assistant Secretary for Energy and Minerals Hollis M. Dole complained so loudly that he was shoved out of the job by Secretary of the Interior Rogers C. B. Morton, who was afraid to press for the Alaska pipeline because of pressure from environmentalists until the OAPEC embargo.

refugees, wondered what the stubble-bearded Yasser Arafat was yakking about at the United Nations, yawned at the news of the dissolution of the Kuwaiti and Bahraini legislatures and the rebellion in Oman, shrugged at the assassination of King Faisal of Saudi Arabia by a deranged nephew, but was intrigued by the movie, *Death of a Princess,* a true love story in which a Saudi princess and her lover paid the death penalty for adultery when they shot her and cut off his head. After the Arab embargo, the big concerns on Main Street, U.S.A., were inflation and oil...which caught the ear of Congress.

There Ought to Be a Law!

The passage of the Economic Stabilization Act of 1970 marked the beginning of a decade of energy legislation. Congress passed bill after bill, year after year, hoping to solve the energy shortage, although there were many who believed the shortage was contrived by Big Oil or the Arabs were to blame. Announcing Project Independence in 1973, President Nixon made a Freudian slip:

> ...by the end of this decade we will have developed the potential to meet our own energy needs without depending on any foreign *enemy* — I mean energy — sources.

Bull! However, Congress continued through Presidents Ford and Carter to squander billions of dollars. To name but a few laws, there was the Emergency Petroleum Allocation Act of 1973, followed by the Federal Energy Administration Act of 1974 before they really got organized and passed the Federal Energy Reorganization Act of 1974, not to be confused with the Department of Energy Organization Act of 1977. There was also the Energy Policy and Conservation Act of 1975, which many mistook for the Energy Conservation and Production Act of 1976. In theory, the statutes were designed to control oil prices, allocate sales and reorganize the bloated number crunching bureaucracy. Four volumes of regulations churned out by the thousands of bureaucrats to enforce the misguided laws only served to promote inefficiency, *raise* prices and add billions in social costs — in excess of a billion dollars a year in administrative and government reporting costs alone. As an afterthought, in case the oil companies made too much money after the regulations were phased out, Carter added the Crude Oil Windfall Profits Tax in 1980.

The above does not include two pieces of legislation *raising* natural gas prices in 1977 and 1978, which created over twenty confusing price categories. There was also a myriad of multi-billion dollar research programs and tax incentives for using alternate forms of energy. Noteworthy were: the Energy Supply and Environmental Coordination Act of 1974, which really advocated the burning of dirty, polluting coal, but the Congress didn't want mention it and shake up the environmentalists; the Solar Energy Research and Development Act of 1974, encouraging people to move to sunny climates and heat their Jacuzzi with solar panels; the National Energy Conservation Policy Act of 1978, telling Americans to turn down the thermostat and drive small Japanese cars; the Biomass Energy and Alcohol Fuels Act of 1979, urging the burning of garbage and corn; and the Wind Energy Improvements Act of 1980, which would have been successful if it harnessed the hot air blowing from Capitol Hill.[75]

"What has this got to do with the Mid East and OPEC?" Sheik Yamani asked United States officials why should Saudi Arabia fight for reasonable prices when any college economics student could see that the United States price controls were contributing to rising oil prices, worsening the inflation and helping raise interest rates to 21 1/2% (the same rate the Mafia loan sharks charged). The soft-spoken Yamani was too much of a gentleman to point out that Congress was panicking and dipping into the pork barrel with each piece of legislation.[76]

---

[75] Details of the legislative folly may be found in *Federal Regulation of Energy* by Professor William F. Fox, Jr., and *Energy Law* by Professors Donald N. Zillman and Laurence H. Lattman.

[76] The suave sophisticated Ahmed Zaki Yamani was a graduate of New York University Law School and Harvard Law School, where he studied international law.

President Carter, naive to Washington's ways and the congressional pork barrel, was astounded to discover many members of Congress were more interested in helping their constituents and campaign contributors and getting reelected. Carter declared his "moral equivalent of war" on the energy crisis from the White House, bundled in a cardigan sweater next to a roaring fire — symbolic of turning down the thermostat to save energy.

THE GREATEST GIVEAWAY AWARD

The prize for the biggest handout was awarded to the Energy Policy and Conservation Act of 1975, which subsidized small oil refiners with an "entitlement" of approximately $1.80 a barrel, coincidentally the price of foreign oil five years earlier, so they could compete with lower cost large refiners. Inefficient refineries with capacities of 10,000 barrels a day or less — derogatorily called "teakettles" by Big Oil — sprung up to take advantage of the government's philanthropy, paid by the major oil refiners who passed the cost on to the consumers. After the boondoggle was discovered by the Department of Energy and the entitlements disappeared, so did the uneconomic teakettles but, while they operated, each teakettle boosted the price of oil and sapped $540,000 out of the economy and into their pockets every month.

Unable to squeeze 10,000 barrels a day out of their ramshackle refineries, many teakettles contracted to have their oil "processed" by larger refiners and split the $540,000 a month with the processor. The monthly cost of the legalized fraud on the consumers was estimated in excess of $25 million.

Congress did some good, but it came too late, proving the theory that Congress doesn't pass acts, it reacts. It finally passed the Trans-Alaska Pipeline Act of 1973 and the Outer Continental Shelf Lands Act Amendments of 1978. The forty-eight inch, 789 mile pipeline from

Alaska's North Slope oil was given the okay to be built after the 1973 embargo, but construction would not be completed until 1977 — eleven years after the oil bonanza was discovered.

Federal oil leasing in the deep waters of the Outer Continental Shelf (OCS), delayed after an oil spill in Santa Barbara Channel, was ordered quadrupled to cover 10 million acres by Nixon. Carter opened up another 55 million acres (about the size of Utah). Not to be outdone, Jim Watt, Reagan's first Secretary of the Interior tried to offer the entire OCS for drilling, but no one took his deliriums seriously except a few environmental groups who declared him "Nature's Enemy Number One."

Oil exploration in the OCS waters is expensive, time consuming and risky. Environmental suits filed under NEPA often delayed lease development in the OCS oil for years and, in some areas, prohibited drilling. British Petroleum's American subsidiary, Standard Oil Company of Ohio, drilled an exploration well in the Beaufort Sea off the coast of Alaska at Mukluk. The result was a $2 billion dry hole.

It is contended by many oilmen, if the Trans-Alaska Pipeline had proceeded on schedule and drilling off the coast of California wasn't delayed because of the 1969 oil spill in the Santa Barbara Channel, the United States could have thumbed its nose at the Arab embargo.[77] The contention is debatable, but everyone agrees that 2 million barrels a day flowing from Alaska — 25% of America's production — would have helped ease the pain.

---

[77] The oil industry's argument may be found in *Fiasco* by Jack Anderson with James Boyd, and *Crisis in the Oil Patch* by Donald P. Hodel and Robert Deitz. Hodel was Secretary of Energy and the Secretary of the Interior under President Reagan.

"Participation" or "Nationalization" — What's in the Name? We Want it All

While Libya, Iraq, Algeria, Nigeria and Venezuela were nationalizing the oil companies, several the Arabian peninsula nations were more subtle. They wanted to participate in the companies' profits. The OPEC resolution passed a few years earlier demanding 25% of the companies' stock was outdated. Taxes were increased from 55% to 85%, cutting the companies' gross profit margins in some cases to 50¢ a barrel. Not satisfied, Kuwait demanded 60% of BP's and Gulf's Kuwait Oil Company in January 1974, then took the remaining 40% in March 1975 and told BP and Gulf to get lost. BP's and Gulf's demand for $2 billion compensation for their investment was sneered at. They were paid $50 million and told to take it or leave it.

The Saudis, less greedy and always thinking of the future, had no choice but to take over 100% of Aramco by the end of 1974 or they would look like American stooges to their Arab neighbors. Yamani negotiated a relatively fair bargain based on the book value of Aramco's assets. The big difference was Aramco was paid to continue operating the oil fields and allowed to market 75% of Saudi Arabia's production for many years.

By the end of 1975, all the Mid Eastern governments owned and controlled their oil reserves.

THE BLUE-EYED ARABS

British Petroleum discovered oil in the North Sea in 1970. Shell and Exxon followed by finding an elephant-sized North Sea field. After the great expense and risk were borne by the oil companies in one of the most treacherous oceans in the world to drill for oil, Britain formed a state oil company, British National Oil Company, and legislated the right to take 51% of the oil production through a tax. Following the Oklahoma based Phillips Petroleum Company 1969 strike in Norway's North Sea waters, its state oil company, Statoil, went into the oil business and controlled production and prices.

The "Blue-eyed Arabs," as they were called by the oil companies, had learned the OPEC game of participation. Today, Britain ranks twentieth in world oil reserves and ninth in production, Norway is fourteenth in reserves and sixth in production.

The United Nations and the Rule of Law

Prelaw students should be jumping up and down and shouting: "What the hell is going on? How about the sanctity of contracts between the companies and the OPEC nations? Some concessions were to last until the year 2000 and the price agreements were to last five years!" Law students may quote Latin: "Pacta sunt servanda" or, more likely, "Sue the bastards! Remember when that banana-nosed weirdo, Mossadegh, took over AIOC? The AIOC seized their British oil when it arrived in European ports." Economics majors will no doubt clamor: "OPEC is a cartel! It's a violation of the American antitrust laws." ROTC members might grit their teeth and swear: "If the Arabs use oil as a weapon of war, they're asking for it. We've got a bigger army and our air force has smart bombs!"

The arguments were settled in the United Nations in December 1974 by the lopsided vote of 110 to 6 with the passage of the Charter of Economic Rights and Duties of States. The Charter put the finishing touches on the 1966 UN resolution calling for Third World nations to grab back their natural resources the colonial powers had controlled for decades. The United States and Britain voted against the resolution. Many developed nations abstained, which they did regularly when they didn't have enough oil and were afraid to offend the oil exporting nations. The trouble was there were a hell of a lot more "developing" nations, who wanted a free ride on the capital and technology of the rich industrialized countries, than "developed" nations. Most in favor of the resolution refused to admit they were poor and called themselves the "Group of 77" instead of "Third World countries."

Article 2 of the UN Charter gave Britain and the United States heartburn. It let the world know that nationalization and expropriation were the in thing to do. There was also a tricky loophole providing: "appropriate compensation should be paid by the State...taking into account its relevant laws...and all *circumstances* the State determines pertinent." The favorite circumstance conjured up for not paying the oil companies was that the original oil concession was granted by a corrupt government. As all the shahs, amirs and sheiks who signed the concessions were greedy corrupt bastards and pocketed the cash, it worked every time. (There were wags who claimed that Shah Muhammad Pahlavi squirreled away $3 billion.) Many nations also argued that they had been raped by the oil companies when they controlled the oil market, played hanky-panky with the oil prices and chintzed on the royalties, so the oil companies should get nothing.

All nations didn't skin the oil companies. Saudi Arabia was farsighted and wanted to stay on the good side of Uncle Sam, who was selling it F-14 jets and other shiny armaments. Ibn Saud's sons didn't trust their neighbors, the power-hungry Shah and the manic-depressive Saddam Hussein, and enjoyed the comfort of knowing America had a bigger and better navy and air force than the two trouble makers. Also, being a kingdom left them open to communism being spread by Russia. The princes were aware that communists love to overthrow monarchies, especially if the royal family is pocketing the oil riches belonging to the proletariat.

The UN Charter also blessed OPEC by recognizing the right of nations to form cartels to develop their natural resources and fix prices, and warned that it would be dirty pool for the powerful industrialized nations to retaliate with economic or political sanctions against the poor nations in the cartel who were just trying to make a buck.

America's favorite weapon — lawyers — failed. When the American worker was squeezed by the high oil prices, the International Association of Machinists Union sued OPEC for violating American antitrust laws. The judge tossed the case out, citing the "act of state" doctrine. This esoteric-sounding legalese says in effect that foreign nations and OPEC can do anything they want and, even if the judge ruled against OPEC, the Arabs could tell him to stick his gavel in his ear.

The "Libyan safety net," set up by the concession holders to share in any cutbacks in production or losses caused by the prima donna Qaddafi grabbing their oil fields, was held unenforceable by the courts under the act of state doctrine. When Bunker Hunt, the son of oil billionaire H.L. Hunt, went to collect from the Seven Sisters after Qaddafi nationalized his oil concession, the Sisters said: "Sorry, it was an act of state."

Americans, who want to be loved by everyone, were shocked when Qaddafi seized American properties in Libya and ranted, "We proclaim loudly that this United States needs to be given a big hard blow in the Arab world on its cold, insolent face..." Qaddafi also warned he wasn't concerned about American capitalistic aggressors invading. The Libyan people would destroy the oil fields and go back to riding camels, as they had wandered around the desert on the smelly beasts long before they had Fords.

IT JUST WASN'T ABOUT OIL

India demanded American businesses turn over their technology to local companies. One of the first to pull out of India was IBM, closely followed by Coca Cola, who didn't want Pepsi Cola to learn its secret formula.

The Solemn Declaration...or Share the Wealth

It didn't take long for Third World countries to realize that they couldn't afford to pay the same high price for oil as the industrialized nations. The increase in the price of oil exceeded the total foreign aid being doled out by the wealthy industrial powers annually and Third World economies were going down the tubes.

A few OPEC nations with extra cash laying around established development banks to loan money to their friends.[78] Most funds came

---

[78] Generally, loans are not made to enemies unless one want to buy them off, which happens to be an Arab custom. One of the largest banks was the Islamic Development Fund, capitalized at $2.4 billion (Christians need not apply).

from Saudi Arabia, Kuwait and the UAE, with relatively small populations and enormous wads of petrodollars sitting in the bank. More heavily populated Iran and Iraq were spending their cash squelching dissent and building up their military forces. Venezuela claimed it was neither a Third World nation nor a first class capitalistic power and admitted it was second rate by saying, *"No mi problema."* Other OPEC members were developing nations, trying to play catch-up by buying shoes and television sets. They also said Third World nations never pay back loans and it was capitalistic to loan money.

Sniping at the OPEC nations made rich by oil became a favorite sport of poor countries and the Arab left wing rabble. Finally, OPEC decided to convene a meeting in Algiers in March 1975 under the grandiose title: Conference of Sovereigns and Heads of State of OPEC Member Countries. Over one thousand delegates and hangers-on arrived with their hands out, eyes glazed at the thought of an Arab socialist society and mouths watering at the mention of a potential $15 billion development fund.

Conference rhetoric was pumped up with timeworn left wing phrases: "new economic order," "dialogue on cooperation" and "solidarity between OPEC member countries in the face of aggression." During a dull moment, they chastised the consuming nations for using too much oil, particularly the "capitalistic Americans." But, in the end, all that was agreed was a *Solemn Declaration* promising "economic cooperation and coordination."

The reason it was called a "Solemn Declaration" was to remind the wealthier nations to keep a straight face when talking to the have-nots. The socialists thought the Solemn Declaration was a bust — they didn't obtain money or subsidized bureaucracies charged with doling out billions in grants. One lefty Arab dabster looking for a handout whined: "OPEC itself turned its back on the Algiers' philosophy and crawled back into its corporatist shell."

What really happened was the crafty kings and sheiks of the Arabian peninsula — the ones with the big bucks — decided to parcel out the money themselves to more friendly faces and not leave it in the hands of a gang of communists and socialists who might want to toss them out on their royal keisters if they got power. The Shah agreed with them for another reason. He needed to buy three more squadrons of F-15s from the United States because Saddam Hussein was buying MiGs from the Russians.

OPEC, which has meetings to decide when to have meetings and meetings to decide on the agenda for the next meeting, met three more times during 1975. This did not include two consultive meetings when the oil ministers sneaked off to the bright lights of Vienna to get away from their wives. (The Muslim members can have up to four wives. If you know how one wife can bang your ear, imagine what four can do.)

The June meeting was a bust. Is was held in dusty Libreville, Gabon, not known as a swinging town. All they could think of to do was admit Gabon as a full member and resolve never to return.

The September meeting was productive. OPEC raised oil prices "by only ten percent," according to OPEC's official chronology.[79] While not mentioned in the official minutes, this rankled their Third World friends, who were still waiting for OPEC's Solemn Declaration to pump money into their treasuries. The poor nations, like Rwanda, Somalia and Sudan, knew they were paying seven times as much for oil than they were two years earlier. OPEC members climaxed the meeting by congratulating Venezuela for nationalizing its oil industry.

The December meeting in Vienna was a blast, well, almost a blast. Carlos the Jackal showed up with bombs. His real name was Illitch Ramirez Sanchez; however, he thought, if he was to become the world's most renown terrorist, "Carlos the Jackal" would frighten more people. Born in Venezuela, the son of a communist lawyer, he was affiliated with the Popular Front for the Liberation of Palestine (PFLP), but had been known to terrorize people and governments on a freelance basis.[80]

---

[79] *OPEC General Information & Chronology.* OPEC Secretariat (1994).

[80] His father gave him a Russian name to honor Lenin and packed him off to Patrice Lumumba University in Moscow to study economics. Apparently, he also took night courses in bomb making.

Minutes before the OPEC ministers broke for lunch on the second day, Carlos stormed into the meeting and announced: "We have launched a campaign of political contestation and information against the alliance between American imperialism and the reactionary, capitalistic Arab forces..." To prove he was serious, Saudi Arabia's Ahmed Zaki Yamani and the Iranian Oil Minister, Jamchid Amouzegar, were condemned to die.

Carlos' next order of business was to separate the delegates and staffs into three groups. The first group were called "Neutrals," and included the non-Mid Eastern representatives from Ecuador, Indonesia, Nigeria, Venezuela and the new guy from Gabon. Next, he picked out the "Friendlies," Libya, Algeria, Kuwait and Iraq, who funded his PFLP pals. Last, Carlos pushed the delegations of the "Enemies" towards the middle of the room: Saudi Arabia, Iran, UAR and little Qatar.

After setting explosives and ordering one of his men to sit in the middle of the Enemies with his finger on a button, Carlos let the Neutrals go to their hotels rooms to watch the fiasco on television. When he counted heads, he flew into a rage. The oil ministers from Qatar and the UAE were missing! Rumor was they had become bored with Vienna and hopped a plane to London to check the action in the discos. The Friendly oil ministers soon discovered that being a friend didn't count for much. Carlos gathered up the ministers from Algeria, Kuwait, Iraq, Iran, Libya and Saudi Arabia, plus enough staff members to fill the seats of a DC-9, and ordered the pilot to fly to Algiers where he had arranged a friendly reception.

Carlos spent a few hours in Algiers before the TV cameras demanding that oil be used to help poor Arabs and Third World countries and making snide remarks that Yasser Arafat and the PLO were too wishy-washy and conservative, then ordered the plane to take off for Tripoli, Libya, another friendly spot on his itinerary. However, the pilot told him the plane had mechanical problems and mumbled a few other standard airline excuses for the bad service, so they headed for Tunis, Tunisia. The Tunisians, never ones to get involved in anything controversial, turned out the lights and parked trucks on the runway, forcing the plane back to Algiers. When the Algerians had enough of Carlos' ranting, they told him they were coming in with guns blazing like Sylvester Stallone in a *Rambo* flick.

The newspaper accounts are a garble of conflicting disinformation from this point. OPEC reported simply: "The meeting was interrupted by terrorists and no decisions were taken." Either a deal was made or Carlos turned wimp. (Rumor was he was promised $5 million.) At the end of the thirty-six hour ordeal, Carlos and his band were allowed to leave Algeria for friendly, sunny Libya.

No one knows who set up the terrorist attack. Yamani claimed it was Qaddafi. Before Carlos left, he promised to kill Yamani and Amouzegar the next time he saw them. Not surprising, the Libyan and Algerian ministers had been released en route earlier, leaving Yamani, Amouzegar and the Iraqi and Kuwaiti ministers to sweat out the entire journey. When Yamani finally got home to Saudi Arabia, he was still caressing his ever present worry beads. King Khalid gave him a Rolls Royce for almost being a martyr.

When the January 1976 meeting rolled around, you can guess what happened. OPEC established the OPEC Special Fund for Third World countries with an initial contribution of $800 million. Half the fund went to the International Fund for Agricultural Development; *provided, the developed countries kicked in $600 million.* (Fat chance.) The $800 million was called a "showcase" by the Arab socialists and a "drop in the bucket" by American taxpayers used to shelling out billions in foreign aid every year.

At the December 1976 meeting, OPEC members added another $800 million to the Special Fund, then announced a 10% increase in the price of oil and scheduled another 5% price hike in six months. To show their spunk, Saudi Arabia and the UAE announced they would only raise prices 5%. Yamani swore that his reason for the smaller increase was because prices were already too high and he could sell more oil if his price was cheaper, not because he was afraid of Carlos. The UAE minister had no comment because he always agreed with Yamani and didn't want to explain to his wives why he left the December meeting early.

THE OPEC SPECIAL FUND

The original $800 million contribution came from: Iran $210, Saudi Arabia $202, Venezuela $112, Kuwait $72, Nigeria $52, Iraq $40, Libya $40, UAE $33, Algeria $20 and Qatar $18.

For those who don't have a calculator handy, the above adds up to $799 million. Little Gabon kicked in the last token million. The members' contributions amounted to about one day's oil production. Ecuador claimed it didn't have any spare change and its daily output was less than a couple of Saudi wells. Indonesia, with most of its population living in poverty and its state oil company, Pertamina, bankrupt because of corruption and incompetence, couldn't cough up a dime.

Hero or Traitor? — Peace or Treason?

Anwar Sadat realized that Kissinger had double-crossed him and had no intention of forcing Israel to withdraw to the 1967 borders or agree to a Palestinian state. Although the United States dumped more aid into Egypt than it could spend, its real purpose was to keep the Soviets out of the Mid East. Arab hopes that the 1976 election of Jimmy Carter, a dedicated humanitarian, would change things were shattered after Carter and Brezhnev reached an accord on an Arab-Israeli peace settlement. Israel and the American Zionists attacked Carter for being pro Palestinian and acting without Israel's approval. This surprised Yasser Arafat, whose PLO was still denied recognition by the United States. Things looked glum for Sadat when the hard-nosed, former terrorist, Menachem Begin, and the Likud Party came to power in Israel in March 1977.

The loose cannon, Sadat, took things in his own hands and went to Israel. An amazed world watched on television as a dignified Sadat offered an olive branch to Begin before the Israeli Knesset. Carter jumped in the love feast by inviting Sadat and Begin to the United States where, after twelve days of tough bargaining, they signed the Camp David Accords in September 1978. Not only did they agree on a peace settlement between Egypt and Israel, they set a framework for a Palestinian solution. Sadat, aware of the danger of sticking his nose in the Palestinian question, threatened to walk out, but was begged by Carter to agree to sign separate agreements. In March 1979, with much political fanfare, Carter witnessed Sadat and Begin sign the peace treaty and Sadat's death warrant.

The agreements were models of ambiguity. Begin took Sadat to the cleaners. Carter sweetened the deal by promising to support Israel by "diplomatic, economic and military measures" including supplying it oil if its supplies were cut off. Where Carter would have obtained the oil during an embargo he never explained. Naturally, Sadat received his share of promises and cash from Carter. During later disputes over what the Camp David Accords meant, Begin stood like a rock and pointed to the fine print while the naive duo of Carter and Sadat pleaded the "spirit of Camp David."

The Palestinians, not invited to the settlement and photo session at the White House, and the PLO, not recognized by Israel and the United States, said: "Up yours! We didn't agree to anything!"

Sadat, adored by the West, was despised as a traitor by the Arab nations, who cut off aid and diplomatic relations, then tossed Egypt out of the Arab League and moved the League's headquarters from Cairo to Tunis. Although America shoveled more aid in Sadat's lap, he lost control of the government, forcing him to arrest 1,500 Egyptian leaders, including Muslim clerics. In October 1981, Sadat was gunned down by Islamic "extremists" according to the West. To the Arabs, Sadat betrayed the Arab world by making a separate peace with Israel and agreeing to surrender Palestine over to the Jews.

Hosni Mubarak, Sadat's Vice President wounded during Sadat's assassination, became President. A pragmatist, he retained a tight control over the government, but freed Sadat's political prisoners, then re-established relations with the Arab world. Mubarak and Egypt were welcomed back into the fold by all his Arab brethren except Syria.

Shah Mat & the Islamic Revolution

The petrodollars failed to trickle down and benefit the Arab in the *souk*. To the peasant, oil revenues meant only inflation. The Arabs had been defeated in 1967 and 1973...disgraced. The promises of socialism had not been kept; a united Arab nation had not come about; and nationalism within arbitrary borders drawn by the West had failed because of corruption and brutality. More and more found the answer in Islam, which provides more than a religion. It is a way of life for man and government. The *umma* — the community of Islam — is the united perfect world.

Things were also going to hell in Iran while Carter was busy pressing the Egyptian and Israeli peace treaty. Shah Muhammad Pahlavi's attempts to westernize Iran failed for many reasons — inflation (the cost of oil caused the cost of F-15s to rise), vast sums spent on a military, overspending on projects that failed to help the masses, brutal suppression of dissent, the growing split between the haves and have-nots and, most important to the average Iranian, the Shah was "un-Islamic."

The opposition of Ayatollah Khomeini and the Shiite clergy to westernization, including land reforms extending to the valuable *waqf* lands controlled by the wealthy clergy, had forced Khomeini's arrest and exile to Iraq.[81] Carter's human rights ideology made him openly hostile towards the Shah, which told the world that the Shah had lost his best friend — America was no longer supporting its former guardian and protector of the Gulf.

The city of Abadan saw the first scenes of violence in August 1978, the most notorious being the burning of a theater showing "Western sin" after over 500 patrons were locked inside. The demonstrations soon reached Tehran, the most westernized city in Iran. Strikes at the Abadan

---

[81] *The Devil's Dictionary* by Ambrose Bierce defines clergyman: "A man who undertakes the management of our spiritual affairs as a method of bettering his temporal ones."

refinery and in the oil fields cut oil production from 6 million barrels a day in August to less than 2.5 million in December, when exports ceased and the employees of the oil consortium were evacuated for their safety.

Saddam Hussein, concerned about Khomeini's rising influence over the Shiite Muslims in Iraq, which made up the majority of Iraq's population, was happy to honor the Shah's request that Khomeini be expelled and shipped off to France in October. It was too little, too late. Khomeini increased his preaching against the Shah from Paris and went on TV, in part, because of the mysterious murder of his eldest son a year earlier. On January 16, 1979, the Shah left Iran on a "vacation," never to return. Two weeks later, Khomeini showed up in Tehran to take over and execute any of the Shah's supporters not smart enough to leave, plus those who didn't agree with his ideology or who had been corrupted by Western influences, particularly the United States — "the Great Satan."

Khomeini proclaimed that all Muslim nations in the Mid East were corrupt, not Islamic enough and should be overthrown, which made his neighbors nervous and think that maybe the Shah wasn't so bad after all. The nation was renamed the Islamic Republic of Iran under the doctrine of *velayat-e faqih* — "government by Islamic judge" — and the Ayatollah Khomeini was the judge over the President, Majlis and everyone else.

Unknown to the State Department and CIA, the Shah had been diagnosed as having cancer in 1974. In September 1979 Carter first refused to permit the Shah to visit the United States for medical treatment, but was pressured by Kissinger and others to allow him to enter the country under an assumed name on October 23, 1979. It was like trying to slip Dolly Parton into a kindergarten without being noticed. The Shah was spotted immediately. On November 4 an Iranian mob stormed the American embassy in Tehran, taking 66 Americans hostages and seizing classified documents. Fifty of the hostages would be held for over one year — 444 days of hell — while America stood helpless.

An impotent Carter could only freeze Iran's assets in the United States and embargo its oil imports to America, which Khomeini would not have allowed anyway. It was too late for the altruistic Carter to learn a basic lesson in diplomacy — it's not always a choice between good and evil — in many cases, it's picking the lesser of two evils. The Shah, compared to Khomeini, was Mr. Nice Guy.[82]

In April 1980 an attempt to free the hostages was aborted in the Iranian desert when two American helicopters malfunctioned and a third crashed into a C-130 and exploded, killing American soldiers. America's military might looked pathetic in the eyes of the Arab world.

Khomeini's war against the Great Satan and un-Islamic nations of the Mid East spread. The American embassies in Libya and Pakistan were attacked by mobs; and embassy staffs in Islamic nations were cut back and emergency evacuation plans made. The Shah had supported Israel, which was to his benefit economically and kept him in favor with the United States. Khomeini reversed the policy and pointed to America and Israel as enemies of Islam. Carter's failure to force Menachem Begin to return to the 1967 borders and overt pro-Israeli positions he claimed all Americans cherished, but few understood, added to Khomeini's arsenal against the Great Satan.[83]

Iranian-inspired Shiites seized the Grand Mosque in Mecca, Saudi Arabia, the holiest of Islamic shrines. The Mecca riots were an affront to Saudi Arabia's sacred duty to protect the Holy Cities and an attempt to drive a wedge between Saudi Arabia and America.

Checkmate — Shah Mat — the Shah is dead. In July 1980 the Shah died in Egypt after being granted asylum by Sadat, which didn't help Sadat's standing with his fellow Islamic Arabs and added another reason for his assassination.

"Both Sides Should Lose"

That was Kissinger's sentiment when Iraq invaded Iran in September 1980. Saddam Hussein understood Khomeini's ranting to

---

[82] The Shah's story is told by his sister, Princess Ashraf Pahlavi, in *Time for Truth*. America is the home to 1.2 million Iranians. Most fled from Khomeini's wrath, but more than a few ran to escape the Shah's depotism.

[83] Jimmy Carter, *The Blood of Abraham*.

Iraq's majority Shiite population across the border to overthrow his regime. The Shiites, with no voice in Iraq's government, had suffered decades of oppression. Saddam was the prime target because he had forced the seventy-six year old Ayatollah to leave Iraq at the Shah's request. He decided to attack Iran, although it held three times Iraq's population. Saddam counted on the Kurds in the north and Arabs in the south of Iran, who were being persecuted by Khomeini, to revolt and the demoralized Iranian military to be ineffectual. The time was ripe to gain support because the entire world was running scared of the weirdo Ayatollah. Claiming Iranian aggression, he denounced the 1975 agreement he made with the Shah allowing Iran access to the Shatt al-Arab waterway[84] and invaded, hoping for a quick victory.

Saddam Hussein miscalculated. Although Iraq had a powerful army by Mid East standards, so did Iran. It rose to the occasion to defend its homeland with the stockpile of modern weapons the United States had supplied the Shah. The army was joined by the *Pasadaran* and *Basij,* volunteers under the Islamic Revolutionary Council, who went to their martyrdom in waves of human cannon fodder. The Iranian army officers had another incentive. When they lost, Khomeini had an estimated 10,000 officers sent to prison or executed.

The war was to last eight years, devastate the economies of both nations and end up in a stalemate, allowing both Khomeini and Saddam to brag that they won.

The Iran-Iraq War also added devastation to the economies of Third World nations and played havoc with the American economy and politics.

Gas Lines & $40 a Barrel Oil

OPEC's meeting in December 1978 was typical. They voted to cover inflation by scheduling four gradual price increases over the next

---

[84] The Shah supported the Kurdish rebels in northeast Iraq in 1974; however, he withdrew his financing and sheltering of the Kurds in 1975 in return for Iraq's recognition of Iran's border being the median line of the Shatt al-Arab, which gave Iran free passage in the outlet to the Persian Gulf needed by its Abadan refinery.

nine months, raising the price from $12.70 to $14.54 by October 1, 1979. The members paid little attention to the first signs of panic spot buying caused by the strikes in Iran, as there was plenty of oil available to make up the loss. After Iranian exports stopped, cutting off British Petroleum's main source of supply, the spot price hit $10.00 over the official OPEC price. British Petroleum and Shell had no choice but to suspend their supply contracts under their *force majeure* provisions. The Saudis cried *"Insh'allah,"* when Khomeini's oil workers were fighting among themselves and unable to get Iran's production on stream, then turned their oil wells wide open, raising Saudi production from 8.5 to 10.5 million barrels a day. Saudi Arabia laughed all the way to the bank as did the four American Aramco members.

QUIZ: What does the term *force majeure* **mean?**

Answer: A legal term taken from the French meaning "superior or irresistible force," excusing a party's performance (delivery) under a contract due to impossibility, such as an act of God, strike, war or event beyond his control. The strike in Iran was a valid excuse on BP's part. Some OPEC nations, such as Nigeria and Indonesia, deemed their contracts with the oil companies were subject to *force majeure.* These nations found it "irresistible" not to honor their contracts because someone else would pay them a higher price for their oil, particularly the Japanese, who wet their pants and pay any exorbitant price when their fuel gauge reads they're a running half empty.

Panic and oil traders drove the price double the OPEC price, often assisted by princes and corrupt national oil company officials for a "commission" — *baksheesh* paid to a Swiss bank account. OPEC couldn't keep track of the skyrocketing spot price. At its March 1979 meeting, it pushed forward the marker price to the October price, *plus anything else the members could get.* Only Saudi Arabia and the UAE held to the $14.54 price. In June OPEC raised the marker price to

$18.00 and decided its members could add "premiums," not to exceed a total of $23.50. Of course, most members cheated.

QUIZ: What is meant by spot price?

Answer: The price paid for the delivery of a cargo at market price on a specific day, as distinguished from a contract price for delivery over an agreed period. Prior to 1979, 90% of international oil sales were made by contract. Since the oil panic of 1979-81, the majority of oil sold is based on the spot price, and contracts are based on published spot market prices.

The blue-eyed Arabs, Britain and Norway, raised their prices along with OPEC and everyone else. (Who wouldn't?) *American domestic oil prices increased under complex price control regulations.* Carter, waking up to the fact that price controls don't work and inflated prices, announced the phasing out of the controls, which outraged the liberals who were blaming everything on Big Oil, so he tempered it with a windfall profits tax to freeze oil company profits.

When Iraq invaded Iran in September 1980, over 5 million barrels of oil for export were locked up by the war,[85] and prices went through the roof. The OPEC meeting in December set the marker price at $36.00 and a maximum of $41.00. (You guessed it, spot prices were reported to be as high as $45.00.) Only Saudi Arabia, Kuwait and the UAE sold below the marker price and pled for sanity in the runaway market that was devastating the world's economy.

---

[85] Experts disagree how much oil production was lost. Prior to the war, Iran produced roughly 6 million barrels a day and Iraq 3 million. The production of both nations varied from during the war, particularly after they started bombing each other's terminals and Syria cut Iraq's pipeline. Regardless of the exact amount lost, the uncertainty caused panic in the oil market during the war.

Carter saw his reelection hopes drown in oil and the faces of American hostages. Joe Sixpack was sitting in long lines and paying the highest gasoline prices in history (when he could find a gas station open on an odd or even numbered day depending on his car's license number); Joe's retired parents living on a pension couldn't pay their heating oil bills; every night, Joe and his wife saw American hostages in Iran on TV; and Mr. and Mrs. Sixpack were unable to borrow money to buy their dream house at 21% interest.

The American hostages would not be released until January 20, 1981. It was Iran's final kick in the ass to ex-President Carter on the day of Reagan's inaugural.

AMERICAN HOSTAGES

Gary Sick, a staff member of the National Security Council under Carter and Reagan, claimed Reagan campaign officials negotiated a secret deal with Iran to delay the release of the hostages until after the election. However, the Democrat-controlled House Foreign Affairs Committee concluded there was no evidence of such a dastardly deal...Sick's accusations were sick. Gary Sick, *October Surprise: America's Hostages and the Election of Ronald Reagan.*

20

FROM PEANUT FARMER TO ACTOR —
Impotence to Blazing Guns

Ronald Reagan, the "Great Communicator" and former B movie actor, was in for a rude awakening if he thought being Governor of California — The Land of Fruits and Nuts — was a tough job. As governor and President, Reagan seldom avoided confrontation. He was also savvy enough to pick a good staff made up of experienced officials. Philip C. Habib, whose parents were born in Lebanon, was appointed his special envoy to negotiate a peace in Lebanon. The appointment of Habib, the first American of Arab ancestry to represent America in the Arab world was welcomed by the Arabs, but wasn't enough to offset the United States' votes favoring Israel in the United Nations to erase many Arabs' hatred of America — "ally of the Israeli murderers of innocent women and children."

In May 1981 Habib obtained a cease-fire in Lebanon but, as Arafat was pulling out his PLO, Menachem Begin double-crossed him and unleashed air raids killing PLO members and civilians. Reagan telephoned Begin to demand a stop the "needless destruction and bloodshed," which enabled Habib, with Saudi help, to arrange a second cease-fire with Begin, Arafat and Assad of Syria. The war was too complex to cover in college World History — 101; so you will have to read about its horrors described in *The History of Lebanon 100 1/2*.

THE HISTORY OF LEBANON — 100 1/2

The enigma called Lebanon is smaller than Connecticut. Its people are a hodgepodge as a result of being conquered by everyone passing by — Babylonians, Phoenicians, Greeks, Arabs, Crusaders and Ottoman Turks, before the French grabbed an area known as Mount Lebanon or *le Petit Libon* in 1920, then added a chunk of Greater Syria and called it le *Grand Libon.*

Mount Lebanon was a natural refuge for many persecuted minorities, notably Christian Maronites and Druze, an Islamic sect, who lived in relative harmony until the Turks started beating up on the Maronites and France screwed things up by adding just enough Muslims from Syrian to maintain a slight Christian majority.

"Majority" was euphemistic. The Christians had their own majority, the Catholic Maronites, plus more minority sects than a coon dog has fleas. The Muslims — Sunnis, Shiites and Druze — didn't get along either. To make it appear everyone was represented and guarantee that the government wouldn't work, the French provided that the President must be a Maronite Christian, the Prime Minister a Sunni Muslim and the Speaker of Parliament a Shiite Muslim.

After the founding of the Syrian Protestant College in 1866 by Americans, later renamed the American University of Beirut, the university became the center of free thought in the Mid East without the usual overbearing missionary zeal to convert its students (but enough hype to keep contributions rolling in from New England Protestants for Bibles printed in Arabic). Lebanon achieved the highest literacy rate in the Arab world. Besides becoming the banking center of the Mid East, it evolved into the playground of the Eastern Mediterranean, offering skiing in the mountains, sun bathing in the warm waters, luxury casinos and hotels, including a Holiday Inn for American tastes.

Lebanon boiled over in 1958 when the Christian President, Camille Chamoun, opposed Nasser's Arab nationalism and insisted he serve another term contrary to the constitution, which split the Christians and Muslims and angered many Christians. The hostilities ended quickly when Eisenhower sent American Marines and a compromise was reached.

Palestinian refugees flooding the country after the First and Second Arab-Israeli Wars became an economic and social burden. When King Hussein kicked the PLO out of Jordan in 1970, adding more refugees, the Lebanese made the mistake of allowing the PLO to run the squalid refugee camps, which soon became bases for Yasser Arafat's PLO to attack Israel. Israel struck back with air raids and armed the Christians, who were fighting the Muslims and PLO. Afraid of Israel's direct intervention and the potential expansion of its borders, Syria sent troops to support the Christians in 1975, confusing the oddsmakers by putting Syria on the side of Israel. The warring factions succeeded only in killing thousands and destroying most of Beirut before reaching a tenuous peace. The truce vanished after the United Nations peacekeeping force pulled out and the militias were rearmed. The second phase of the civil war saw Syria shift to the side of the Muslims and PLO against the Christians and Israel.

Habib's 1981 cease-fire was followed by Israel's June 1982 invasion, paraded as a defensive measure called "Peace for Galilee," and occupation of a 25 mile zone into Lebanon to protect Israel from PLO rockets. Israel's initial pretext for the invasion was the wounding of the Israeli ambassador in London, but when it came to light the assailant was a member of the anti-PLO Abu Nidal, Israel was forced to change it's excuse for invading. The war turned into a carnage by ruthless barbarians on *all sides:* the slaughter of civilians, the assassination of Lebanese leaders, the bombing of the American embassy in Beirut killing 46, the suicide bombing of the peacekeeping force in which 241 American Marines and 58 French paratroopers were killed, and the massacre of over 1,000 Palestinian civilians in refugee camps by Christian militia while Israeli soldiers watched with approval. The horror was enough to cause demonstrations by Israelis in Tel Aviv against the butchery.

America's part in the war only added to the confusion. Reagan ordered the bombing of Syrian missile sites after American reconnaissance plans were fired upon by the Syrians. However, it was not enough to overcome Syria's influence and the setback in American prestige in the eyes of the Arabs stemming from Reagan's withdrawal of the Marines after the suicide bombing of the Marine barracks, which shocked Americans and caused them to ask: What the hell are we doing in Lebanon?

The baffling 16 year civil war, intensified by Israeli and Syrian self-interests, lasted until 1990, and involved battles between Druze and Shiite "militias" (politically correct term for murdering gangs, including the Christian militias) until the Syrians took charge and backed a new Lebanese Muslim-controlled government with its mechanized army and money.

Remember, Syria has always believed that Lebanon was part of "Greater Syria."[86]

◆ ◆ ◆

---

[86] An excellent reporter's view of the Lebanese War, entitled "A Nation Lay Dying," may be found in six-time Pulitzer Prize nominee David Lamb's *The Arabs: Journey Beyond the Mirage.* The candor of Caspar Weinberger, Reagan's Secretary of Defense, in *Fighting for Peace* covers the Mid East during the Reagan years and is fascinating reading from an incisive insider's view.

Typical Sad Obituary
Malcolm Kerr, President of the American University of Beirut was murdered by two members of the Islamic Jihad on January 18, 1984. Tens of thousands of Muslims and Christians united in their mourning for the educator, peace-maker and true friend, then went back to war.

The wars raging in Lebanon and between Iraq and Iran scared the *Jahannam* out of the Arabs living on the *al-Jazirah*. Having the hell scared out of the nations living on the Arabian peninsula, resulted in Saudi Arabia, Kuwait, Bahrain, Qatar, Oman and the UAE forming a mutual defense pact called the Gulf Cooperation Council in May 1981. Yemen, the only other *al-Jazirah* nation, was a Marxist *Iblis* (devil), and wasn't invited to join.

America was also regarded as an *Iblis* in June when the Senate blocked the sale of AWACs and F-15s to Saudi Arabia under lobbying pressure of the American Israel Public Affairs Committee, despite Reagan's pleading. It wasn't until Anwar Sadat was assassinated in October 1981 did the Senate approve the sale by a narrow margin under fear that the United States might lose Egypt as an ally.

During the June 1982 invasion of Lebanon, the United States was the only nation in the 15 member UN Security Council to abstain from voting for a resolution calling on Israel to lift its siege of the war torn land. OAPEC nations discussed another oil embargo on the United States, but decided against it because the shortage of oil had eased and the Arabs were funding Iraq's war against Iran with oil revenues.[87] Another reason for Arab caution was that judicious King Fahd had just succeeded the weak and ailing King Khalid in Saudi Arabia and he wanted to take delivery of the balance of the F-15s and AWACs and determine which direction Reagan would lead America.

---

[87] Syria, the weird kid in the neighborhood who always hangs around on the wrong side of the street, supported Iran as did the bad boy of North Africa, Qaddafi.

In June 1981, when Israel bombed an Iraqi nuclear reactor outside of Baghdad, Reagan applauded. Menachem Begin claimed Iraq was about to complete its first nuclear reactor, but his opponent, Shimon Peres, said it was a lot of electioneering *bopkess* (goat shit). Few Westerners believed Begin, but didn't complain because they didn't like the idea of a maniac, like Saddam Hussein, running around with a nuclear bomb except the French who were helping him build it. The Arabs, who knew Saddam better than the Westerners, weren't too crazy about it either.

Two weeks after Reagan's 1984 re-election, the United States resumed diplomatic relations with Iraq. After urinating at each other since the Six-day War in 1967, America extended the hand of friendship to Iraq, then counted its fingers and checked for missing rings, typical of the lesser-of-two-evils diplomatic doctrine. To the Arabs, it appeared Reagan was backing them against their foe, Iran. However, it was difficult for King Fahd to figure out what America was up to because there were a couple of wackos in Reagan's National Security Council (NSC) brain trust — Admiral John Poindexter and Lt. Colonel Oliver North — and Fahd wasn't too sure about William Casey, the head of the CIA who had become his buddy. No history of the Mid East is complete without a few words on the Iran-Contra affair, sometimes called "Irangate" or "Ollie's Follies."

Ollie's Follies

The Iran-Contra story starts in 1984 when a hick Senator from Oklahoma named Boland offered an amendment to the Defense Department appropriations prohibiting military funding or support in the Nicaraguan civil war, which the White House got in a twit over because they were supporting the Nicaraguan rebels, the Contras. It was a simple case of a liberal Democrat Senate telling a conservative Republican President they didn't want the right wing Contras overthrowing the left wing Sandinista government of Nicaragua. The NSC came up with a plan Reagan later said no one told him about. The Democrats said Reagan lied. The Republicans retorted that the plan was too complicated for him to understand or he fell asleep during the meeting.

The NSC plan called for secretly selling missiles to Iran and using the profits to fund the Nicaraguan Contras. In theory, this would get the United States on the good side of the Iranians, incline them to become more "moderate" and help obtain the release of American hostages taken in Lebanon and hidden in Iran. To assist in the sneaky work, Ollie hired the Israelis to deliver the missiles because of their experience in sneaking arms into Iran in its war against Iraq. For a while, they made so much money, they also secretly funded the Afghan *mujahedeen* fighting the Russians. Middlemen, Manucher Ghorbanifar, a former member of the Shah's SAVAK (nasty secret police), and Adnan Khashoggi, a Saudi wheeler-dealer, made a few million *baksheesh* on the deal.

In October 1986 a typical braggadocio soldier of fortune pilot, Gene Hasenfus ("chicken or coward" in German slang), crashed in Nicaragua and spilled his guts about Ollie's Follies when captured by the Sandanistas. Reagan fired Ollie and Poindexter; and Congress and the press had a field day at the Iran-Contra Senate hearings.

It was never determined if the NSC conspirators were smoking something illegal during their planning sessions, but it was obvious that they broke a lot of laws, including lying to the Senate. Ollie's boss, National Security Advisor Robert McFarlane, pled guilty. A jury convicted Ollie, but the verdict was overturned on a legal technicality. Ollie insisted he was innocent and a great patriot when he ran for the U.S. Senate in Virginia as a Republican. After losing badly in an election year when the Republicans scored their biggest congressional victories in modern history, Ollie became a talk show host and a darling of the right wing.[88]

The Iran-Contra affair caused the Arabs to ponder what the Jahannam America's foreign policy was.

---

[88] Virginia Senator John Warner, a staunch Republican conservative and one of the Senate's most experienced defense experts, showed his courage by refusing so support North's candidacy as did everyone in the Reagan White House who knew Ollie.

Reagan & the Mid East Bullies

Reagan wasn't a pussycat, and wouldn't take any crap from Libya's Qaddafi, who he called "the mad dog of the Middle East." After Qaddafi claimed that Libya owned half of the Mediterranean, which he labeled "The Zone of Death" for all trespassers, the United States Navy shot down two Libyan jets and sank two armed scows Qaddafi called a navy. When Qaddafi's terrorists bombed a Berlin discothäque, killing and injuring American servicemen, Reagan ordered the Air Force to bomb the hell out of Tripoli and Benghazi. Qaddafi, aware the bombs were aimed at his keister, crawled into a hole and kept his mouth shut for several years.

America's friends reacted differently. The Iron Maiden, Britain's Prime Minister Maggie Thatcher, cheered her friend's quick draw and blazing guns. France, dependent on oil from North Africa, refused to permit the Americans to overfly French territory on the bombing run and publicly said the bombing was a *faux pas*. In private, French President Francois Mitterand told Reagan he hoped the bombing wouldn't be a "pinprick."

Oil finally dragged the United States into the Iran-Iraq War. Oil runs the machines of war and generates the money to pay for war. Syria's closure of Iraq's pipeline to the Mediterranean and Iranian attacks on Iraqi oil installations were offset by the money pouring in from Saudi Arabia, UAE and Kuwait. When Iraqi missiles started hitting Iran's offshore oil facilities and tankers, Iran struck back at Kuwait's tankers and lobbed a few missiles at Kuwait to let them know they didn't like them sticking their nose in the war. In 1986 Kuwait asked the United States and the Soviets to protect their tanker fleet; however, Reagan told Kuwait that it was either an all-American show or nothing. He didn't want the Russians to make any friends in the Mid East. Avoiding getting legally involved in the war was handled by hanging the Stars and Stripes on the Kuwaiti tankers and the United States Navy escorting the "American" ships through the Persian Gulf.

QUIZ: <u>You</u> should ask: I thought Kuwait was a pal. We went to war to save its ass from Iraq in 1991. Why did Kuwait ask Russia to help them?

Answer: Kuwait was on friendly terms with the Soviets because the United States refused to sell arms to the tiny sheikdom for fear it would upset the balance of power in the Mid East and the weapons might get into the hands of anti-Israeli terrorists. Naturally, Kuwait bought weapons from the Soviets. (American foreign policy does not always include economics or foresight.)

Everyone got into the act to protect their oil supplies — Britain, France, Italy, the Netherlands and little Belgium. West Germany sent ships to the eastern Mediterranean to replace the American fleet sent to the Gulf. Russia had to be satisfied with sending four tankers flying the Soviet flag and daring Iran to attack them. Japan, barred by its constitution from going to war, promised to send money. In April 1988 a desperate and foolhardy Iran attacked an American naval ship. The result was quick and deadly. The United States Navy sank six Iranian naval ships and blasted the hell out of several Iranian offshore oil installations.

On July 3, 1988, the *USS Vincennes* shot down a plane the pilot foolishly failed to identify. Unfortunately, it was an Iranian commercial airliner and all 290 civilians aboard were killed. Khomeini felt like a harlot in a mosque. No major powers strongly criticized America, and he was feeling America's powerful wrath. Earlier, Iraqi jets had mistakenly attacked the *USS Stark* and gotten away with it by saying, "Oops, I'm sorry." Reagan blamed Iraq's missile attack on the Soviets and Iran for causing the confusion. Also, the world shrugged when Iraq used poison gas on Iranian troops. The Russians, who were selling arms to both sides, gave Khomeini little comfort. When Iran captured a Soviet freighter carrying arms to Iraq, it was forced to release the ship and cargo when several Soviet warships headed towards Iran with orders to shoot.

On July 17 Khomeini agreed to a cease-fire, mumbling that he would have preferred to drink poison and someday he'd get even with Saddam Hussein, the Saudis and the American Satan. He didn't. He died a few months later.

There are a lot of Iranians who still hate America's guts...but then they hate everyone who is not as perfect as they are.

STOP!   I didn't say that all Iranians hate Americans. That's like saying everyone in Washington is a Redskin fan and hates the Dallas Cowboys. (Only 99.44% of the Washingtonians wear "DALLAS SUCKS" buttons.) Politicians and historians too frequently generalize because it's easy to describe another nation in sweeping terms as "American Satan" or "Bomb-throwing religious rag heads." In Iran, politicians and the *mullah* clergy are often one in the same because the Iranians are smart enough not to elect lawyers like we do in America. Iranians believe Americans are nogoodniks for many reasons, including our backing of Shah Muhammad Pahlavi, who most everyone now agrees was a rotten scoundrel, and *gharbzadegi,* loosely translated to mean "Westoxification" — leading the pure-minded Muslims into the Western world of sin, much like many Baptists believe the Unitarians are going straight to hell.

Oversimplified, all the United States has to do to gain Iran's friendship is admit that the Shah was a SOB, promise not to spread sin in the Mid East, return the billions in frozen Iranian assets held since 1979, allow Iran to build nuclear bombs and tell the Israelis to pack their bags and go back where they came from...Well, at least it's a starting point for negotiating a peace worthy of a Nobel Peace Prize, assuming we can find the "moderates" in Iran Ollie North was hoping to meet and bribe. A good place to start is with Iranians who believe the powerful clergy — the "Mercedes Mullahs" — are backward and corrupt. Many Iranians claim that we can also bribe the *Mullahs*.[89]

Of course, everyone knows what a "moderate" is. The press and politicians use the term to describe individuals and entire nations every day. Former Senator Barry Goldwater, now revered as an elder statesman, responded to being labeled an extremist in his acceptance speech at the Republican presidential nominating convention in 1964: "Extremism in the defense of liberty is no vice...and...moderation in the pursuit of justice is no virtue!" One never admits to being an extremist, but may claim to be a liberal, conservative or moderate (somewhere in the middle). The other guys are the damn extremists! A moderate is someone who is not as extreme as the screwballs who oppose your viewpoint. Extremist nations become moderate when they need allies in a war or money to bail them out of trouble. Conversely, when America needs allies in a war or *oil*, we see moderation in nations we once thought extreme. Reagan thought Iraq was more moderate than Iran.

As Reagan's second term came to an end, the savage Lebanese war remained heated; American hostages continued to be taken and murdered, including the CIA's chief of station in Lebanon; and the UN was still debating the "Palestinian question." Over two million Palestinian refugees had fled their native land and another 945,000 were living like prisoners in Israel, the West Bank and Arab lands annexed by Israel after

------

[89] For an insight into what the average Iranian is thinking today, I recommend the unlikely title: *Know Thine Enemy* by former CIA case officer, Edward Shirley.

the 1967 war.[90] Every peace plan failed, primarily because of Israeli rejection, including the Fahd Plan in August 1981, the Reagan Plan in September 1982, the Arabs' Fez Plan of September 1982 and the Shultz (Reagan's first Secretary of State) Plan of February 1988.

In December 1988 Arafat appeared before the United Nations to accept UN resolutions 242 and 338 on behalf of the PLO, including Israel's right to exist. However, Israel's Yitzhak Shamir denied the right of an Arab Palestinian state to exist and refused to recognize the pre-1967 war boundaries. The tables were turned. Now it was Israel demanding the removal of Arabs from Palestine, and the hell with the 1948 UN partition plan establishing "Independent Arab and Jewish States." The Israelis were teed off because Reagan recognized the existence of the PLO and the bearded little schlemiel, Arafat.

King Fahd and other Arab leaders, frustrated by American seesaw policy, asked why the United States didn't force its will on the bully, Israel, after Johnathan Pollard was caught spying for Israel against America.

QUIZ: Who is King Fahd Ibn Abdul Aziz al-Saud and what is he up to?

Answer: King Fahd is the pro-American son of Ibn Saud. As Crown Prince, he evidenced his support of the United States by going into a self-imposed exile over his brothers' vehement refusal to bend to Carter's pressure to support the Egyptian-Israeli peace in 1979. To understand Fahd and the al-Sauds, one must recognize the dangerous tight-rope Saudi Arabia

---

[90] According to the United Nations Relief and Works Agency, the Diaspora, a name originally given to the Jews driven from Babylonia, had scattered Palestinian refugees around the world. Per the 1988 UN census, they were primarily spread over Jordan 1,150,000, Lebanon 490,000, Kuwait 300,000 Syria 230,000 and Saudi Arabia 140,000. America welcomed 100,000 displaced Palestinians, half of whom became taxicab drivers in Washington, D.C.

walks in the Mid East. Being the richest nation in the Mid East has its drawbacks. Its radical neighbors with larger populations and armed forces run by Islamic fundamentalists, leftists opposed to a monarchy and just plain crazies, such as Iran, Iraq and Libya make it a prime target. The plight of the Palestinians make it difficult for any Arab nation to appear "moderate" towards Israel. (There's that word again.)

The Saudis must strike a delicate balance between the United States, who they depend on for their defense, against maintaining a tough stance against Israel and the anger of their Arab neighbors towards America because of its support of Israel and its flaunting of Western un-Islamic ideas. At home, as well as with his fellow Arab leaders, King Fahd lost face when he had to grovel for F-14s and missiles before a pro-Israel American Congress. When the United States stood by and watched its best friend in the Mid East, the Shah, fall to Khomeini, an avowed enemy of the West and Saudi Arabia, Fahd was concerned whether he could depend on the flighty Americans.

Saudi Arabia's internal problems — dissidents demanding democracy, Islamic fundamentalism, the curtailment of services and national projects because of low oil prices, coupled with vast expenditures for defense — continue to cause headaches. During the oil boom, Saudi Arabia influenced the Arab world by financing the Arab-Israeli wars, buying friends and paying off enemies — spending hundreds of millions it no longer can afford. Notwithstanding, many still believe it is a bottomless money pit because of its immense oil riches.

The al-Saud brothers run Saudi Arabia by consensus but that does not mean they are united or fully trust each other. The army and air force are under Fahd's full brother, Minister of Defense and Aviation Prince Sultan; the National Guard was under his half-brother, Crown Prince Abdullah; and the Ministry of Interior, which houses the police, intelligence and Coast Guard, is controlled by his full brother, Prince Naif;

thus, reducing the chance of a coup.[91] Crown Prince Abdullah, the next in line for the throne, was opposed to the close ties to the United States. Some say he has mellowed and changed his thinking since the Gulf War, but it must be kept in mind that pro-American King Fahd is 77 years old and has suffered a stroke. As a result, the more austere Abdullah has taken over the duties of running the nation.

QUIZ: What is Islamic "fundamentalism?"

Answer: In the West, it's a catchword, depending on who is using the phrase. To the American press its the Islamic version of "born-again Christians" with beards hellbent on a literal reading of the Koran instead of the Bible. If an Arab blows up the World Trade Center or a Palestinian wraps himself in explosives and kills everyone on a bus in Tel Aviv, including himself, he may be a member of *Hezbollah* (Party of Allah) or *Hamas* (Islamic Resistance Movement), and we will curse him as a fundamentalist on a *jihad* (holy war) seeking to become a *shahid* (martyr).

To the young female Saudi Vassar graduate, fundamentalists are those who ban her wearing a Gucci dress and lipstick and driving a Mercedes when she is home. Her brother, the MIT engineer or graduate of the London School of Economics, cringes at the uneducated, pious fundamentalists setting back progress and free thought. Saudi Arabia is a bas-

---

[91] This is not mere paranoia. It's the way things are done in the Mid East, intra-family and military coups are standard fare. Kings and dictators don't run for election. Saddam Hussein will not allow his generals and cabinet members, some of whom are related, to carry weapons in his presence. In 1969 when a Nasser-inspired coup was uncovered against the al-Saud, rumors flew that the coup leaders were taken for a one-way plane ride over the Empty Quarter, although several culprits later showed up in jail.

tion of Islamic fundamentalism, whose strict Wahhabi Islam history is traced back to 1744 when Ibn Saud's great-great-great-grandfather, Muhammad Ibn Saud, adopted the teachings of Muhammad Ibn Abdul Wahhab. Today, the al-Sauds attempt to take from the West only the technology and knowledge they believe compatible with Islam. Like the "pro-life" and "pro-choice" groups in America, Muslims disagree on what is acceptable behavior. As part of his political agenda, the ultra-fundamentalist Khomeini, called Saudi Arabia corrupt and materialistic and instigated the takeover of the Grand Mosque in Mecca to protest the Saudis being "un-Islamic."

An Islamic nation is fundamentalist if its laws are derived solely from the *Sharia* (Islamic Law). A cardinal fundamentalist tenet is to protect the purity of Islamic precepts from adulteration (e.g. Western thought). Islam is a way of life that transcends oil riches, as Sheik Ahmad Zaki Yamani noted in 1981: "The Holy Cities of Mecca and Medina are on the other side of Arabia from the oil fields. But they are part of the same country — and in our eyes they matter more than anything else."

Even when a half million American troops were defending Saudi soil against Saddam Hussein, King Fahd insisted that General Schwarzkopf's army not offend fundamentalist Islam. (No booze.)

There is no doubt that King Fahd has the fundamentalists on his mind when he deals with the United States. Yet, to many Americans, King Fahd is a "fundamentalist." Crown Prince Abdullah, King Fahd's apparent successor, disdains the lavish lifestyle of his half-brothers, and only time will tell if he is more of a "fundamentalist" and whether the 73 year-old will change Saudi Arabia's policies towards Israel and the United States.

The Law of Supply & Demand Can Be a Harsh Law

The uncertainty of access to oil supplies at the start of the Iran-Iraq War shot oil prices over $40 a barrel, and "experts" were predicting "$50 oil" was just around the corner. With 20/20 hindsight, it is easy to understand what happened. The demand for oil was dropping in 1979. In America, the world's biggest oil glutton, consumption declined from 18.3 to 16.5 million barrels a day between 1978 and 1980, almost 10%, because of conservation measures and the simple fact that Joe Sixpack couldn't afford to fill his tank as often. Gradually, new giant oil fields in Alaska and the North Sea and numerous smaller discoveries added to the supply. Marginal oil fields uneconomic at $3.00 a barrel in 1973 could make a healthy profit at $40. Oil at $30 to $40 in 1970s an 1980s was worth taking a bigger risk for in the deep waters of the North Sea, Mexican Gulf and jungles of Malaysia than $1.80 a barrel oil in 1970. The economics of oil exploration had changed. *In ten years, the price of oil had increased twentyfold.*

When the shortage subsided, so did the price. One of Reagan's first acts was to eliminate the price controls and inefficient regulatory allocation of oil Carter had started to phase out. During 1982 and 1983 OPEC tried to act like a real cartel for the first time by setting quotas to limit production and cutting posted prices from $36.00 to $29.00. But quotas were difficult to enforce. Iran and Iraq needed oil revenues to fight a war as did Saudi Arabia and Kuwait who were helping finance Iraq's war. Many members cheated by producing over their quotas and discounting the price. Nigeria, the perennial freeloader and cheater, had no choice because of a costly coup in December 1983. Venezuela, which helped found OPEC in 1960 because it desired a quota system to compete with cheap Mid East oil, said, "Screw it, we need the money" (it sounds better in Spanish), and refused to obey the quota system. Cooperation by the blue-eyed Arabs and our "good neighbor," Mexico, to cut back production to maintain an inflated price couldn't stem the falling prices, so they cut prices.

The big loser was Saudi Arabia. In March 1983 the Saudis agreed to be OPEC's "swing producer" by lowering its production to supply the balance of the market demand, if the other members would honor lower quotas. Between 1981 and 1985, Saudi oil revenues dropped from $119 to $26 billion a year and its production dwindled from 8.5 to 2.4 million barrels a day.

In June 1985 King Fahd warned other OPEC members that Saudi Arabia could no longer tolerate their cheating. Natural gas produced with Saudi oil dropped to the point where it threatened the nation's domestic gas needs. The Aramco partners told Yamani it was impossible to pay Saudi Arabia the full posted price because of the cheating. Yamani and Aramco developed "netback pricing" whereby the price was determined by what the refined products would earn in the market, allowing the company and Saudi Arabia to split the profits and guaranteeing the company a $2.00 a barrel profit (less than 5¢ a gallon). Other producing countries had no choice but to follow suit and continue to cut prices.

In November 1985 the Saudis, disgusted at the fraud on OPEC quotas, turned their oil wells on full blast and sold as much as they could at whatever the market paid. At the outset, West Texas Intermediate was selling for $31.75. Within four months the American benchmark price plummeted below $10.00 a barrel. *In the Mid East the price dropped to $8.00 a barrel.*

American banks financing the upward spiral of oil prices went into the red. Continental Illinois, the seventh largest bank in the United States, went under, requiring a bail-out of over $13 billion in capital and Federal government loans. Mexico's government, borrowing and spending on the promise of $40.00 a barrel oil, owed America's largest banks $80 billion it couldn't repay, threatening their viability and requiring the United States government to help bail-out the Mexican government and the banks. Small banks in the oil producing states went bust because of defaults on the gamble of oil loans. It also signified the end of many small independent oil companies who couldn't weather the price collapse, especially those with marginal oil wells that were no longer profitable.

Financial institutions and markets teetered all over the world. In Kuwait, the *Suq al-Manakh,* a quasi legal stock market, crashed leaving billions of worthless paper. The pseudo market was like playing with

Monopoly money, often postdated checks carrying 100% interest, until some wise-ass tried to cash one and turned the *Suq al-Manakh* into a worthless mass of sheep dung. The Amir of Kuwait was forced to suspend the National Assembly, which turned nasty after the Amir was accused of bailing out his relatives and friends.

Rock bottom oil prices made a lot of nations unhappy. Non-OPEC oil exporters with diverse interests, such as China, Mexico, Norway, Colombia (which has more than cocaine to sell) and the Soviet Union, attended OPEC meetings. The States of Texas and Alaska showed up to see if they could help OPEC shore up the price. Royalties and taxes paid to the State of Texas were crucial — each $1.00 drop in the price of oil meant between $90 to $100 million a year less going into the State's treasury. Although the Russians didn't want to get involved with a bunch of screaming Arabs in case they were caught cheating on the quota (the Soviets were also selling arms to both Iraq and Iran), they mumbled that they would contribute to the cause by cutting back their exports 100,000 barrels a day.

In April 1986 Vice President George Bush, the former Connecticut Yankee turned Texas oilman, during a visit to Saudi Arabia, urged King Fahd to stabilize the oil market, hinting that Reagan might invoke a tariff if he didn't. Bush caught hell when he returned. Reagan didn't believe in tariffs and said his Vice President was acting like an oilman and pissing off the voters by calling for higher oil prices. Bush antagonized the White House, but not the economists, who saw the domestic oil industry and banks on the brink of disaster with oil at $10.00 a barrel. During the next few months, a consensus developed in the industrialized world that oil at $18.00 a barrel would strike a nice balance; that is, keep the bankers and oil companies solvent.

One night in October 1986, Yamani heard over television that he had been fired. At the next OPEC meeting in November, Saudi Arabia accepted a quota of 4.5 million barrels a day if OPEC would accept a quota of 15.8 million barrels a day, down 31% from OPEC's production of 22.9 million barrels a day in 1980.

For the next four years, OPEC passed flowery overly optimistic resolutions and adjusted quotas in an attempt to reach the $18.00 goal, but the free market price hovered between $15.00 to $16.00 until August 1990 when hell broke loose in the Mid East again.

Is That What Really Happened or Was it a CIA Plot?

There are those in the Reagan White House who claimed King Fahd's decision to increase production from 2.4 to over 8.5 million barrels a day in December 1985 was masterminded by William Casey, the flamboyant Director of the CIA. [92]

The National Security Council plotters knew Reagan was hot to defeat the *Evil Empire* — the Soviet Union. Unnoticed by Congress and many political pundits, who were blaming the Arabs for running up the price of oil, the Soviets were a major beneficiary of high oil prices. They were earning hard currency to the tune of $2 to $3 billion a month to fuel the Cold War by exporting oil at the same price as the Arabs. The NSC knew Soviet rubles were worthless and the Russkies needed American capitalistic dollars and the best way to beat them was economically...bankrupt the bastards!

The NSC formulated a plan to flush the Soviet economy down the toilet by financing the Afghan rebels, supporting the Solidarity movement in Poland, and developing a Strategic Defense Initiative — dubbed "Star Wars," which the Soviets couldn't afford, plus *cutting the price of oil.*

The sneaky Reds were also planning a 3,600 mile natural gas pipeline from Urengoi, Siberia, to Western Europe that would bring Moscow another $2.5 billion a month. Strategically, the NSC believed a pipeline making Europe dependent on the Soviets for over 30% of it natural gas needs during the Cold War would allow the Soviets to blackmail Europe — threaten to freeze them by cutting off gas supplies. Britain, in financial doldrums, and France, always out to make a franc, refused stop their industries from exporting American licensed oil and

---

[92] The tale is told in *Victory: The Reagan Administration's Secret Strategy That Hastened the Collapse of the Soviet Union* by Peter Scheweizer. Read it with beer and pretzels so you can take it with a few grains of salt.

gas technology needed by Russia to complete the pipeline, making it more difficult for the United States to cripple the already vulnerable Soviet economy. Irate, Reagan ordered a stop on the licensing of oil and gas technology.

The CIA's Bill Casey and Defense Secretary Caspar Weinberger developed Operation Peace Shield with King Fahd, the forerunner of Desert Shield and Desert Storm. In 1985 the White House gained the ability to supply additional sophisticated weapons for the Saudis through a pro-Israel Senate, thanks to the orchestration of political virtuoso, Republican Majority Leader Bob Dole, and built up the 300,000 U.S. Central Command (USCENTCOM) that was to be General Schwarzkopf's command in a few years.[93]

Six months after the development of Peace Shield, NSC insiders claim that Casey told King Fahd high oil prices were strangling the United States economically, making it difficult for America's to support the defense of his kingdom, and Saudi Arabia was losing revenues by cutting oil production. Casey advised Fahd that the only real winners were Fahd's enemies, the Soviets, Iran and Libya, who were benefitting from Saudi Arabia single-handedly propping up the high prices. Enlightened, King Fahd began flooding the world with oil...*Bullshit!* Closer to the truth is Defense Secretary Weinberger's analysis: "It was an internal Saudi decision to increase production and cause the price of oil to drop in 1985. But it was a decision that they knew would sit very well with the United States."

---

[93] The Reagan administration was miffed when Fahd refused to wait for the Congress' blessing to allow Saudi Arabia to purchase additional F-15s and bought British Tornado fighter-bombers. But that was nothing compared to the Congress' apoplexy in 1988 when it discovered the Saudis had purchased medium range surface-to-surface missiles from China. King Fahd, aware that the Congress would never approve missiles that could upset Israel's balance of power, didn't bother asking America.

King Fahd and Yamani knew what they were doing — shaking out the cheating riffraff. The Saudis have always favored reasonable prices and exhibited concern about the world's economy and inflation. They also knew they needed the United States to defend the oil fields. Yamani, although extremely competent, never made a major move without clearing it with King Fahd or Kings Feisal and Khalid before him. But no one could foresee how low prices would drop and the difficulty of raising them again. That's why Yamani was unceremoniously canned as oil minister. *The King can do no wrong because he's the King.*

THE NYMEX

One of the major contributors to the free market controlling world oil prices was the New York Mercantile Exchange (NYMEX) opening futures trading on the spot crude oil market in 1983. The public could now gamble on volatile crude oil prices as the nicknames of other NYMEX petroleum futures indicated — Russian Roulette (gas oil), Boston Bingo (heating oil) and Manhattan Mogas (gasoline). It was a way for doctors with inside tips to lose their homes and Cadillacs in a hurry. At first, Big Oil screamed in terror (a few laughed) at "paper barrels" that might have an affect on their market — they dealt in "wet barrels." But, within a year, Big Oil became the major player in the NYMEX, hedging its purchase price of oil from the Mid East, which took months to reach America's refineries, and reducing their risk.

The NYMEX eventually helped stabilize the price of oil, except for occasional blips in the price when Saddam Hussein raged against a small Gulf nation, Khomeini swore vengeance on Kuwaiti tankers or rumors cropped up of another potential coup in Nigeria or Algeria. Speculation before OPEC meetings as to what action they will take still causes fluctuations until the rhetoric dies down.

21

THE BUSH LEAGUE YEARS — Winning the War and Losing the Peace... And Slick Willie Slips in the Slippery Sands

George Bush began his administration with a confusing carrot and stick approach to Israel. First, he warned Israel it couldn't hold onto the lands it annexed during the myriad of Arab-Israeli Wars by building homes for Jewish settlers, then he gave Israel $100 million to stockpile weapons in case of a crisis. Next, he barred the immigration of Russian Jews into the United States, leaving them no place to go except the Occupied Territories.

The Palestinian *Intifada,* started by primarily young Palestinians throwing stones and Molotov cocktails was in full swing. The *Intifada* (shivering or shaking off, like a dog with fleas...Israeli fleas) had erupted spontaneously in December 1987 out of the frustration and humiliation of the people, who saw the leadership of the PLO, now exiled in far off Tunis, unable to help them. Rioting spread through the West Bank and Gaza Strip. The PLO added their support by organizing the secular United National Leadership of the Uprising (UNLP) and the Islamic Bloc (Muslim Brotherhood) formed Hamas.

The *Intifada,* which lasted until 1993, convinced Israel that it was futile to continue the oppression and denial of a Palestinian national identity, forcing it to begin the Oslo Accords negotiations. *Intifada* was not only Hamas terrorism, but effective strikes, boycotts of Israeli products, resignations from government positions and refusal to pay Israeli taxes. The civil disobedience resulted in the death of an estimated 1,700 Palestinians and the arrest of 90,000. During the rising up of the people they exposed Israel's spy network and punished 20,000 Palestinian informers.

Lebanon was so screwed up, Bush stayed out of it. The Amal Shiites, supported by Syria, and Hezbollah, the bomb-throwing "Party of Allah" Shiites backed by Khomeini, fought each other in the streets, while the Christians were trying to run Syria out of the country. Things got so hot, the Lebanese Christian and Muslim parliament had to leave town and meet in Taif, Saudi Arabia, in October 1989 to sign a peace accord and elect a president, who was promptly assassinated when he returned home. A pro-Syrian, Elias Hrawi, was elected in his place, but he couldn't take control until two rival Christian militias stopped killing each other and the Syrians knocked off the winner and drove the militias out of Beirut towards the south. Then the Israelis started shooting back at the Hezbollah thugs over the border and rearming the Christian Southern Army. By 1990 Bush wasn't paying much attention to what was going on in Lebanon because it didn't have any oil and several Mid Eastern nuts were causing trouble.

In Iran, the Ayatollah Khomeini had a fit over Salman Rushdie's book, *The Satanic Verses,* and issued a *fatwa* (religious edict) ordering his death and the destruction of the publisher, Viking Press, to let the world know how he felt about blasphemous prose. A scared Rushdie remains in hiding in England after Khomeini's death because of a $2.5 million price on his head, which doesn't say much for free speech or tolerance in Iran.

Hashemi Rafsanjani took over as President in Iran, but that didn't mean that he ran things. Another Ayatollah named Khamenei showed up with signs saying he was really in charge: *Obedience to Khamenei is Obedience to Imam — Khomeini.* Rafsanjani was replaced by a "relative moderate," who the *Washington Post* spells "Khatemi" and the *Wall Street Journal* dubs "Khatami." Due to the confusion in names and who was the real leader, high school teachers dropped Iran from Current World History — 103.

"Friendship is to be found between individuals, but between nations interests prevail." [94]

When America supported Iraq in its war against Iran, we called our friend "President Saddam Hussein." After he used poison gas on Iraqi Kurdish citizens and bragged that his role model was Joseph Stalin, he became "Dictator, Saddam Hussein." The fact that a nation of 18 million maintained an army of one million malcontents two years after an eight year war that cost Iraq a half a million casualties made his Arab neighbors suspicious. They knew that any leader, who shoots his political opponents and hangs their bodies on meat hooks in public for potential rivals to see what happens to those holding contrary views, isn't a nice guy...he certainly isn't a "moderate."

Some psychiatrists say Saddam's problems stem from being orphaned as an impressionable young boy and being raised by an crazy uncle who wrote a pamphlet: *Three Whom Allah Should Not Have Created: Persians, Jews and Flies.*

Iraq was broke. It couldn't pay back the $30 billion loaned by Saudi Arabia, Kuwait and the UAE during the war. On July 16, 1990, Saddam sent Kuwait a $2.4 billion bill for oil he accused it of stealing from the Rumalyah oil field straddling the Iraqi and Kuwaiti border and said he was cancelling Iraq's debt to Kuwait of $13 billion because Kuwait and the UAE had been cheating on their OPEC quotas, driving down oil prices. During the next few days, Saddam announced that Kuwait and the UAE were in cahoots with the capitalistic United States

---

[94] *Desert Warrior,* General Khaled bin Sultan. Prince bin Sultan, General Schwarzkopf's Joint Commander during the Gulf War, is the first member of the House of Saud to publish a book. *Desert Warrior* underscores the cultural differences and Islamic dogma dividing Saudi Arabia from the West even when America was saving their royal butts. As the Gulf War was also Saudi Arabia's war (they paid for it), it is suggested reading for those desiring an understanding of Saudi-American relations.

to keep oil prices down and threatened to use force to make the Arab exporters maintain their quota, then sent a crew to start drilling for oil in Kuwaiti territory.

The United States, as usual, was little help in the Mid East. In fact, we made matters worse by our customary diplomatic fawning to tyrants. Adding to the debacle was the State Department's affirmative action program. It sent April C. Glaspie as the first female ambassador to an Arab nation, believing Iraq would be a nice place to train the career bureaucrat for an important post. The idiocy brings to mind the Arab proverb: "He that would be buggered supplies his own grease." [95]

On July 25, Ambassador April had her first one-on-one meeting with Saddam after two years as Ambassador. April dutifully sat listening to the blowhard scream about Kuwait's theft of oil and his threat to use force, then replied: "We [America] have no opinion on the Arab-Arab conflicts, like your border disagreement with Kuwait," then went on her annual vacation, like any good bureaucrat. In other words, April told Saddam that America didn't care, it was okay to beat the hell out of little Kuwait.

Meanwhile, the Arab nations tried an "Arab solution," which means lots of talk and *baksheesh,* but no action. After the talks broke down, the Amir of Kuwait complained to the United Nations, then looked out the window and saw 100,000 Iraqi gangsters at his border.

On August 2 Iraq rolled over the nation about the size of Hawaii, sending Kuwait's Amir scurrying to Saudi Arabia with Kuwait's tiny army close behind. When Bush heard the news, he was shocked, but mumbled that he wasn't thinking about intervention. Fortunately for Kuwait and the world, Britain's Prime Minister, Margaret "The Iron Maiden" Thatcher, was visiting Bush and told him: "This is no time to go wobbly, George." Bush found his backbone and condemned Saddam's aggression, then pushed for

---

[95] The Adventurers of April in Arabland are skillfully depicted by Robert D. Kaplan in *The Arabists, The Romance of an American Elite.* It is clear that the career woman, who had clawed her way to the top in the bureaucracy and looked like a missionary frump, wasn't cut out for the job as Ambassador to Iraq and dealing with Saddam.

a UN Security Council Resolution 660 demanding Iraq withdraw or sanctions and military force would be employed.

King Fahd asked the United States to help defend Saudi Arabia and kick Iraq out of Kuwait. Much has been written about Secretary of Defense Dick Cheney arriving in Riyadh with satellite photos of Iraqi troops massed on the Saudi border to convince Fahd to allow the United States to intervene. The truth is that King Fahd didn't like the odds (five-to-one according to *Figure 6*) and he had already agreed with his Gulf Cooperation Council allies that the United States was the only power that could protect their nations from Saddam Hussein...Who else was there?[96]

Figure 6
GULF COOPERATION COUNCIL ARMED FORCES

	Total Troops	Tanks	Aircraft
Saudi Arabia	122,500	550	189
UAE	44,000	206	110
Bahrain	6,000	54	46
Oman	29,500	75	57
Qatar	7,500	24	38
Kuwait	9,600	?	40
Total	219,100	909	480
Iraq	**1,000,000**	**5,600**	**848**

Source: Desert Warrior

---

[96] American peaceniks and a few political pundits cried, "America isn't the world's policeman." In the whimsical view of P.J. O'Rourke in *Give War a Chance:* "But you'll notice that when Kuwait got invaded, nobody called Sweden."

The combined forces of Saudi Arabia's two strongest allies, Egypt and Syria, might make a fight of it, but King Fahd couldn't be sure. Syria was bogged down in Lebanon, and neither were about to leave their borders undefended from Israel. Both were jealous of America's top billing and refused to fight under real general, like Stormin' Norman Schwarzkopf, who might kick ass if they didn't attack when ordered. Egypt and Syria said they would only defend Saudi Arabia and liberate Kuwait, but wouldn't invade Iraq. Also, King Fahd couldn't help thinking that Egypt hadn't won a war since the Pharaohs were running things 3,000 years ago.

Actually, Saudi Arabia didn't know how strong Iraq's military forces were. The United States overestimated Iraq's strength until after its surrender for political reasons — it didn't want to look bad in the unlikely event the scruffy Iraqis shot back and accidently won a battle. Also, it was the first time the Pentagon found someone to pay for a war and the testing of their new smart bombs under actual wartime conditions.

Saddam tried to confuse the Arabs by linking Iraq's withdrawal from Kuwait with Syria pulling out of Lebanon and Israel withdrawing from occupied Palestine, but Bush and Fahd wouldn't buy it. It would have made Saddam a hero to the Arabs and weakened the Coalition. The Israelis said in effect: "There's nothing in the deal for us, and those *goniffs*, Kuwait and Saudi Arabia, have been funding terrorists against Israel."

The United Nations passed a trade embargo against Iraq, cutting off its ability to sell oil. Desperate, Saddam tried to make friends with Iran by agreeing to allow it access to the Shatt al-Arab, which had been his excuse for starting the eight year war. When Saddam heard that the United States and Britain had a few hundred jets ready to bomb the hell out of his military installations, he shipped Western civilian hostages, including women and children, to military bases to act as "human shields." From that point, he was no longer called "Saddam, the dictator," but "Saddam, the cowardly son of a bitch."

Like all SOB dictators, Saddam lied about what he was up to. He claimed Kuwait had always been part of Iraq, and he was merely bringing it back into the motherland. But, everyone knew better. Kuwait's founding went back to 1756 when the al-Sabahs took over — twenty years before the United States told King George it would no longer pay his damn tax on tea.[97] In 1897 Sheik Mubarak The Great signed a treaty on behalf of Kuwait with Britain for protection — long before Churchill's Forty Thieves drew lines in the Ottoman Empire sands in 1920 and created Iraq out of a mishmash of Babylonians, Assyrians, Kurds and Arabs the Brits called Mesopotamians.

An Arab League meeting in Cairo attempted to find an Arab Solution in order to keep the un-Islamic Western powers out of the Mid East, but it only ended Arab "friendships," if there are such things. Of the 20 Arab League nations who showed up, only 12 voted to send troops to support Saudi Arabia, but many had big "ifs" or merely agreed to send some other nation's soldiers. Only the two bigmouths, Qaddafi of Libya and Arafat of the PLO opposed the pleas from Saudi Arabia and Kuwait. Qadaffi mouthed off that Saddam shouldn't have picked on Kuwait because it gave the Arabs a bad name, but he didn't want the United States messing around in Arab affairs. Later, when Qaddafi saw how the vote came out, he offered to send troops, but the Saudis said: "No thanks."

King Hussein of Jordan, sitting next to Saudi Arabia and Iraq and up to his ass in Palestinian refugees, mumbled that it would be neighborly if Iraq pulled out of Kuwait. Mauritania and Sudan, both receiving Iraqi arms, expressed "deep reservations." Algeria, with Islamic fundamentalist riots back home, abstained. Yemen also abstained because it was getting guns from Saddam. Yemen's and Cuba's votes in the United Nations opposing sanctions against Iraq proved they were pinkos, anti-American and anti-monarchist Saudi.

---

[97] The Declaration of Independence also stated that King George III had "erected a multitude of New Offices, and sent hither swarms of Officers to harass our people, and eat out our substance," which Newt Gingrich claims is still going on in Washington.

Tunisia, never one to get involved, stayed home. No one ever heard of Comoros, a couple of tiny volcanic islands off the coast of Africa that makes up the twenty-second member of the Arab League, and probably didn't receive an invitation.

An Arab League scorecard is needed to figure out what really happened.

Figure 7

SAUDI ARABIA v. IRAQ —
ARAB LEAGUE SCORECARD

	VOTE	SCORE *(Troops Sent)*
Algeria	Abstain	Sorry — we have a civil war
Bahrain	Saudi	223 — Coalition base
Djibouti	Saudi	Yes — but we're too small to play
Egypt	Saudi	33,677 — but won't play in Iraq
Iraq	Iraqi	*Lost in 5 days — Mid East Record*
Jordan	Weaseled	Didn't want war in Jordan field
Kuwait	Saudi	9,643 — wanted home field advantage
Lebanon	Saudi	Yes, but army back home fighting
Libya	Iraqi	Saudis didn't want them on team
Mauritania	Weaseled	Really a Saddam fan
Morocco	Saudi	1,327 — 1st to arrive to help
Oman	Saudi	957 — U.S. base during war
PLO	Iraqi	Arafat bets on wrong team
Qatar	Saudi	1,581 — Coalition base
Saudi Arabia	Saudi	95,400 — and paid the bills
Somalia	Saudi	Yes — but we're starving at home
Sudan	Weaseled	Sorry, busy with civil war at home
Syria	Saudi	14,300 — but won't play in Iraq
Tunisia	No show	Never publicly roots for anyone
UAE	Saudi	1,497 — Coalition base
Yemen	Abstain	They lied — they were Iraq fans

Desert Storm – But First Hot Air During Desert Shield

Americans don't like war, especially if the other political party thinks of it first. One hundred and ten Democratic members of the House petitioned Bush to allow more time for the UN sanctions to work. Congress only supported the war by a narrow vote of 52-47 in the Senate and 250-183 in the House — two days before the war was scheduled to begin. Peace marches in San Francisco (where they demonstrate against everything but immortality and sin) made King Fahd nervous. He wondered why he had paid billions of dollars to feed and support a half million American GIs in Saudi Arabia during the last four months and had second thoughts about denying Brooke Shields a visa to visit and entertain. Fahd thought the war was official — Bob Hope had entertained the troops at Christmas as he had done in World War II, Korea and Vietnam.

It wasn't that Bush didn't try to avoid a war. He attempted to arrange a peaceful solution. Bush and Mikhail Gorbachev worked together to broker a peace and warned Saddam that he had to pull out of Kuwait unconditionally. The problem was that Saddam was gambling America wouldn't attack because of the soft Congressional support he saw on C-Span and CNN. Unfortunately, no one took Iraq's Foreign Minister, Tariq Aziz, seriously because he looked like Peter Sellers with a bad case of diarrhea.

Live, Americans watched the war — Operation Desert Storm — start on television with the bombing of Baghdad. While CNN was understandably not allowed to show the live troop movements, the nation saw American men and women prepare for the ground attack, photos of smart bombs being shot down air conditioning vents, Iraqis surrendering to camera crews and the Scuds drop on Riyadh and Tel Aviv. Saddam's obsolete Russian Scud missiles were his most effective weapon because they scared the hell out of everyone. The thought that the madman might have armed the Scuds with chemical weapons and aimed them at civilians caused terror. When Scuds landed in Tel Aviv, Bush feared the coalition would pull apart if the Israelis retaliated. He convinced them to take no action by shipping Patriot antimissile missiles to defend Israel.[98]

The ground war was cut off precisely 100 hours after it started (great White House PR timed for prime time TV). The might and sophisticated weapons of the United States destroyed the highly touted third rate Iraqi army. Much second-guessing has taken place whether the Coalition forces should have disarmed Iraq's Republican Guard and deposed the psychopath, Saddam. It is clear the United States had no desire to kill additional hapless Iraqi soldiers or further risk the lives of Coalition troops, although King Fahd was furious that Bush stopped bombing the Republican Guard in order to eliminate Saddam as a menace. The United Nation's mission was to drive Iraq's army out of Kuwait. The mission was successful.[99] *But...*

Saddam Hussein Got Away With Murder

Generals Schwarzkopf and Khaled bin Sultan went to a small airstrip at Safwan, Iraq, to settle the cease-fire with little or no instructions from Bush and Fahd other than leave the politics to us. They were left to deal with the two low level Iraqi generals they had never heard of instead of an Iraqi leader. Schwarzkopf was "suckered" (his words) when he allowed Iraq to fly armed helicopters after the cease-fire. Schwarzkopf did his job. He was a soldier, not a politician — *and the cease-fire failed to produce a real peace.*

It was wishful thinking by Bush that Saddam would be overthrown by dissident military officers, Shiites who make up over half of the pop-

---

[98] As can be best determined, 88 Scuds were launched: **44 at** Saudi Arabia, 41 at Israel and three at Bahrain. The worst damage was in Dhahran, Saudi Arabia, where a Scud hit a barracks, killing **28** American soldiers and wounding over 100. Hushed up was the news that Israel "ran out" of gas masks before the Palestinians got them.

[99] Accounts of the war are available in General H. **Norman** Schwarzkopf's autobiography, *It Doesn't Take a Hero* and General Khaled bin Sultan's *Desert Warrior.*

ulation and the Kurdish 15% of Iraqi citizens. The Shiites and Kurds heard Bush's mumbling over the airwaves to revolt against Saddam and believed the United States would come to their aid. Eventually, they were massacred by Hussein's Republican Guard and the armed helicopters. Two million Kurdish refugees fled to Iraq's mountains, Turkey and Iran, where many languish in camps and hovels today. Bush and the UN finally created no-fly zones in Iraq to protect the poorly armed rebels. Then Bush went wobbly and announced the American policy of non-intervention in Iraq's internal affairs.[100]

During the war, Saddam ordered Kuwait's oil fields set on fire, resulting in the worst oil spill in history. Millions of barrels of crude oil spewed into the Persian Gulf and the black smoke blotted out the sun for hundreds of miles. The sadistic cowards called the Iraqi army murdered and tortured thousands of Kuwaitis during their occupation and stole everything they could carry when the fled, even toilet paper and Pampers®. Hundreds of Kuwaiti prisoners dragged off to Iraq are still missing.

The United Nations passed a handful of forceful resolutions, demanding Iraq pay reparations and return all stolen property to Kuwait, account for all missing Kuwaitis, destroy its chemical and biological weapons, accept United Nations on-site inspections of the dismantling of weapons of mass destruction, denounce international terrorism, cease the repression of its civilian population (Kurds and Shiites), and recognize Kuwait's boundaries. UN demands were enforced by a trade embargo on all imports and exports except for $2 billion in oil sales for food and medicine. But, after seven years, Saddam has still refused to comply with the sanctions. (The 110 Democrat Congressmen who wanted more time for the sanctions to work prior to starting Desert Storm would have had a long wait.) Saddam rejected selling the embargoed oil permitted under the UN sanctions for humanitarian purposes, if he had to spend the money to feed his people, particularly the Kurds, until May 1996 — 5 years after he got his ass whipped. He retains an

---

[100] "I'll bet Bush would have saved the Kurds and Shiites if they had oil," a tearful American citizen born in Iraq told the author.

iron grip on Iraq through brutality and sneers of defiance at the UN resolutions. Cutting off the ears and branding the foreheads of deserters helps maintain an army he can no longer afford.

Pundits, from the comfort of their ivory towers, predict it is only a matter of time before Saddam Hussein falls to a coup or an assassin's bullet. Meanwhile, Iraq's economy is crumbling and its people are suffering from malnutrition and the lack of adequate medical care.

IRAQ'S OIL-FOR-HUMANITARIAN AID DEAL

The $2 billion oil-for-food-and-medicine deal Saddam agreed to accept from the UN in May 1996, after arguing over how it was to be distributed for five months, was a pittance. Under UN sanctions, every six months Iraq is allowed to sell $2 billion worth of oil (then roughly 700,000 barrels a day) and deposit the funds into an escrow bank controlled by the UN. But first, everyone has to get their cut: 30% for the UN compensation fund to pay for war reparations, 6% for the UN weapons inspection programs and 4% for the UN's oversight of the humanitarian aid ($80 million to the bloated bureaucracy to keep the books and hand out food). This leaves $60.00 for each Iraqi, in the unlikely event there is no graft and corruption, in a nation plagued with malnutrition and lack of basic medical supplies that kills hundreds (some say thousands) of children every month.

In 1993 the madman Saddam (excuse the redundancy) instigated a plot to assassinate ex-President Bush while he was visiting Kuwait, forcing President Clinton to order the bombing of Iraq's Intelligence Headquarters in Baghdad. Slick Willie knew he could be next and warned Saddam that Americans don't like their Presidents getting assassinated, *but it's doubtful if Saddam got the word.*

Clinton faced Saddam's aggression in October 1994 when 80,000 Iraqi troops were massed a few miles from the Kuwait border. The United States shipped 54,000 American troops and 360 aircraft to the Persian Gulf as a show of force before Saddam decided to pull back his army.

What did the United Nations do about Saddam's dangerous gamesmanship? They debated it. Only Britain supported the United States by sending Tornado fighter-bombers. General Khaled bin Sultan was correct in his analysis that "national interests prevail." Saudis are subtle politicians, especially the members of the royal family.[101] While the United States, Britain and members of the Gulf Cooperation Council continue to insist that the embargo remain in effect until Saddam complies with all the sanctions, France, Russia and China desire an early end to the embargo. **You should ask: Why?**

France criticized the United States for overreacting to Iraq's troop movements and was outspoken in its position to lift the embargo, citing the plight of the Iraqi people. It is not a coincidence that French oil companies have negotiated oil concessions on some of the best Iraqi oil fields to take effect after the embargo. The major supplier of weapons to Iraq has been owed $6 billion since the Iran-Iraq war. It isn't a fluke that France will reap most of the profits from the $2 billion to be earned from humanitarian aid.

The Russians, crying that Clinton was being cruel to the poor Iraqi peasants, were signing a $2.5 billion protocol with Iraq to rebuild Iraq's war-damaged oil industry *at the very moment Saddam's army was massing near Kuwait's border* in an attempt to recoup $7 billion owed them by Iraq since the Iran-Iraq war. This should have given Clinton cause to marvel why Russia was asking for investments of billions of dollars from American oil companies, with the United States government's blessing, to rebuild Russia's oil industry that had been decimated by incompetence and total disregard for conservation and the environment in order to obtain hard currency.

---

[101] Prince Khaled bin Sultan's brother, Prince Bandar bin Sultan, has been the inscrutable respected Saudi Ambassador to the United States for over two decades.

China, the third member of the UN Security Council leaning towards lifting the embargo, is waiting to sell Saddam its shiny, new long range missiles.

Prince Khaled bin Sultan was euphemistic in saying "national interests prevail" — he could have said, "national greed prevails." He might also have damned the three nations for contributing to the instability of the Mid East but, as Henry Kissinger found out too late, the Saudis are indirect and subtle in their conduct of foreign policy.

QUIZ: Why did Saddam Hussein move troops to Kuwait's border in October 1994?

Answer: No one really knows. Expert analysis range from a castration complex to theomania. There are those who say Saddam was attempting to deal from a position of strength to get the UN embargo lifted. Others believe he was merely pulling Clinton's chain by only moving his army 100 miles and compelling the United States, Saudi Arabia and Kuwait to spend millions of dollars.

QUIZ: Was General Schwarzkopf correct when he said, "There wasn't enough left of Iraq's army for it to be a regional military threat?"

Answer: If you don't know the answer, ask President Clinton why he sent 54,000 troops and 360 planes to Kuwait in October 1994.

At the end of the Gulf War, America's oil supply was assured...*for the time being.* America's spirits were high. Our military had won a decisive victory over a tyrant. Men and women in uniform had the stigma of the Vietnam syndrome lifted. America's only hurt was the 146 Americans who didn't come home and those disabled. Our hero was General Schwarzkopf, who commanded 540,331 United States men and women and 54,691 troops from 11 other nations. I give you these bland statistics only to impress that it was really an American show.

That is not to say that the Saudis under General Khaled didn't pull their weight and contribute. Thirty-eight Saudi soldiers paid the ultimate price. In addition to 95,400 Saudi troops, Khaled had 89,051 garrisoned from 24 nations under his command. Managing the arrogant French was no easy task by itself, but he also had a melange of languages and cultures ranging from Mid Eastern Arabs to Scandinavians (Norway and Sweden), former European communists (Hungary, Czechoslovakia, Poland and Romania), fellow Muslims (Bangladesh, Afghanistan and Pakistan), Far Easterners (Philippines, Singapore and South Korea), and Africans (Senegal, Niger and little Sierra Leone, which sent a proud contingent of 24). Only a tactful Saudi prince could have managed that assembly.

The Japanese made an *unkept* promise to commit $11 billion to the war to protect their oil interests, *which was what the war was all about.*

Speaking of Oil...

The price jumped $8.27 a barrel on the NYMEX during the three days leading up to Iraq's invasion of Kuwait, from $16.22 to $24.49. The following Monday, it hit $28.05. During Desert Shield, it varied between $30.00 and $40.00. When news of the Desert Storm air strikes was heard early on January 16, the price jumped up roughly $6.50 and hit $40.00. As the oilmen checked the oil prices on their NYMEX computer screens and watched the war on CNN at the same time, they saw how effective our Air Force bombs and Navy Cruise missiles were. The price dropped $8.00 — *closing $1.50 lower than when it opened.*

The following day, thanks to smart bombs, the price flattened faster than an Iraqi soldier diving for cover to $20.00. When the hostilities died down, the oil prices ended slightly lower than they were before Saddam Hussein gave the world heebie-jeebies.

What happened to the oil prices? It was difficult to explain to two "Congresspersons" more intent on getting reelected than determining the truth, as I had the misfortune to attempt. For most politicians, it's easier to blame rising oil prices on Big Oil or Arabs. Actually, the answer is simple. When Iraq invaded Kuwait and the embargo took effect, the oil companies knew that their combined oil production of 4.3 million barrels a day was no longer available to the international market. The companies calculated there was about 3.2 million barrels a day excess capacity outside of Iraq and Kuwait, which meant there would be a worldwide shortfall of 1.1 million barrels. This was supported by the CIA, which believes oil is so important, it keeps track of it.[102]

The big glitch in the math was Saudi Arabia was counted on to contribute 2 million barrels of the excess capacity and Saddam's artillery and tanks were pointing towards the Saudi oil fields. Forget the computations if Iraq destroys the Saudi's oil fields. We could have an 11 million barrel shortage! Whenever there is a shortage on the horizon, prices are going to rise according to the law of supply and demand taught in Basic Economics — 101.

OPEC's Iran, Libya, Venezuela, Nigeria, and the UAE each produced between 200,000 to 400,000 barrels a day more than they pumped prior to Desert Shield. *The United States couldn't squeeze out an extra drop for the war effort!* The big surprise was Saudi Arabia, with its back to the wall and American tanks surrounding its oil fields, spewing out an extra 2.7 million barrels a day. It had to. Desert Storm consumed 40 million barrels of oil. Those jets Americans saw on television devoured 80% of the oil products — 500,000 barrels of jet fuel a day.

When the Congresspersons didn't believe me, I tried to explain the economics in words they could understand...

Two minutes after Joe Sixpack heard TV's Ted Koppel say there might be a shortage of Coors, he drove to the 7-Eleven

---

[102] The CIA calls the Department of the Interior and Department of Energy to check its figures. In turn, the agencies ask the oil companies, who are not about to tell the government how much oil they really have, so the CIA isn't always accurate. (Would Exxon tell Shell?) As you might expect, the Arabs and Iran don't tell anyone how much oil they really have, so CIA's effort is an educated guess. (See Chapter 23 for more about oil statistics and other lies.)

to buy a case, even though he had a few cans in his refrigerator. Abdul, the 7-Eleven manager, also heard the news and that Coors increased its price from 50¢ to 75¢ a can because of the shortage of fine barley and hops in Colorado. Abdul, who had sold Coors at 75¢ a can that morning and sips a cold Coors Light himself when no one's looking, now had to replace each Coors at 75¢. Abdul wasn't about to sell Joe beer at what it would cost him to buy. By the time Joe arrived at the 7-Eleven, Abdul had raised the price of Coors to $1.00. Shocked that the price of Coors might skyrocket, Joe bought two cases instead of one, adding to the shortage.

Some might say that Joe doesn't have to drink a six-pack every night. Economists call that demand elasticity — when the price of beer goes up, some can't afford it and drink less. But not Joe Sixpack, he needs his nightly brewski. Nor is gasoline elastic. Joe has to drive to work everyday. When Abdul looked at the shelves and discovered his 7-Eleven was almost out of Coors, he raised the price to $1.10.

"But it wasn't that serious...no one knew if Saddam Hussein would attack Saudi Arabia," the Congresswoman insisted. "What was the real reason?"

I answered her question with a question: "Would your insurance company increase the rates on your house if it knew that Saddam Hussein had burned your neighbor's house, claimed your kitchen was on his property and had a cannon pointed at your front door?" My point was, if oil companies believe there is a danger of a shortage, they buy more oil, just like Joe Sixpack, driving the price up. She didn't answer, but went on a tirade against insurance companies, her second favorite unscrupulous industry, as she sipped her bottled Evian water. As one whose job was representing oil companies, I had done my best, at least she was no longer attacking oil companies. I swigged my Scotch, thinking her problem was she drank too much Evian water after I noticed what it spelled backwards.

Of course, the oil market is more complex than I explained to our elected leaders. The amount of oil available is effected by the stocks in hand, production costs, transportation, refining capacity, and America's Strategic Petroleum Reserve (SPR). By the time President Bush ordered the token release of 33 million barrels from the SPR to ease the fear of a shortage, the oil companies didn't need it. So much for the Connecticut/Texas oilman's knowledge of the oil industry.

Big Oil took it on the chin when it agreed to Bush's request to voluntarily freeze oil prices at the outset of Desert Storm. When the prices fell, Big Oil was accused of price fixing. As a result, I got indigestion during an exasperating dinner with two naive Congresspersons.

If the Members of Congress weren't constantly running for reelection, I might have admitted that oil companies were in business, which surprises many overpaid "servants of the people". Oil companies, like other businesses, maximize profits whenever possible by *charging what the market will bear.* That's why gasoline prices go up faster than they go down. As sophisticated as Big Oil is, it's not perfect. Sometimes it screws up in the mad scramble for oil and passes its mistakes to the consumer, *if the market will bear it.*

Big Oil has never been loveable, and it's history is less than admirable. Many of Big Oil's actions are hard to defend, some impossible, such as the Achnacarry Agreement. Eugene O'Neill's memorable line from *A Moon for the Misbegotten* summed it up: "Down with all tyrants! God damn Standard Oil!"

During the oil crises of 1973 and 1980, "American" oil companies were *accused* of diverting cargos of Mid East oil bound for the United States to the more profitable European market in Rotterdam. Was this wrong? (No one proved oil was diverted.) Are Exxon, Mobil and Texaco American companies? (Saudi Arabia owns a piece of Texaco's operations.) Is British Petroleum still British? (Kuwait owns 10% of its stock.) Or are these monster-size international corporations merely producing, buying and selling oil wherever they can at the most profitable price? If so, is that wrong? Who are Petroleos de Venezuela, Saudi Arabian Oil Co., National Iranian Oil Co. and Kuwait Petroleum Co.? More important, what are they up to?

Big Oil no longer controls oil prices or production, nor does OPEC. Presently, the free market is the arbiter under the law of supply and demand. However, as the United States and the other nations of the Western world run short of oil, they will become more dependent on OPEC and Mid East oil nations and their national oil companies...And there are a lot of nuts like Saddam Hussein running around the Mid East that can screw up the supply of oil before it reaches Joe Sixpack's local Texaco station.

THEORETICAL ECONOMICS — 100 $^1/_2$

Adam Smith, the Scottish scholar, published *An Inquiry Into the Nature and Causes of the Wealth of Nations in 1776.* Its study is required for students of economics and should be compelled for a liberal arts education and Congresspersons. A few quotations from this wily Scot discussing his conclusion: "The market price of every particular commodity is regulated by the proportion between the quantity actually brought to market, and the demand of those who are willing to pay the natural price of the commodity ..." show he predicted what Big Oil and OPEC would do over two centuries ago:

"It is in the interest of all those who employ their land, labour or stock, in bringing any commodity to market, that the quantity never should exceed the effectual demand; and it is in the interest of all other people that it never should fall short of that demand." (Did Smith foretell market demand prorationing?)

"The monopolists, by keeping the market constantly understocked, by never supplying the effectual demand, sell their commodities much above the natural prices." (Did OPEC read the *Wealth of Nations?*)

Another eye-opener — Adam Smith recognized that there had to be a few extra petrodollars for the kings and sheiks over what is required to run their nations: "Over and above the expenses necessary for enabling the sovereign to perform his several duties, a certain expense is requisite for the support of his dignity...We naturally expect more splendor in the court of a king than in the mansion-house of a doge or burgomaster." This explains the luxurious palaces throughout the Arabian peninsula. The rent-free White House Americans provide their President ain't too shabby either.

Long before Adam Smith, there was recorded a *hadith* (saying) of the Prophet Muhammad: "Only Allah can fix prices." Thus, OPEC's price fixing schemes are *haram,* which means price fixing ain't "kosher."

22

THE SHIFTING SANDS — What the Hell's Going On?

The shifting sands of the desert are not only shaped by *shamals,* violent sandstorms amounting to a *nakba* (catastrophe) such as the Six-day War, but also by political winds. The collapse of the Soviet Union far from the Mid East blew breezes that were felt in the desert and shifted the sands of national interests. Syria, Libya, Iraq and radical Palestinian elements were left without a superpower to fund their bad habits and infuse a false sense of backbone. America's concern of a communist threat diminished, but there is still the problem of Russian, French, Chinese, etc., etc., etc. (everyone) arms sales to the troublemakers for money rather than ideological reasons. Russia assisting Iran in the development of a nuclear reactor and supplying Iraq with chemical weapons capabilities are also ill winds.

Saudi Arabia and its Gulf Cooperation Council friends will suffer many sleepless nights if Iran obtains enough enriched uranium for a nuclear bomb. New Yorkers should also stock up on Nembutal and Miltown — the next time someone tries to blow up the World Trade Center, the bomb might also take out the Brooklyn Bridge and Shea Stadium. The only guys in the neighborhood with the *chutz-pah* to stop the madness are the Israelis, who Americans secretly hope will send a couple of planes to blow the hell out of the Iranian reactor as they did when Saddam Hussein started building a nuclear reactor.

In the coming years, the dangerous games will be terrorism and assuring that Iran, Iraq, etc., etc., etc. (no one) acquires weapons of mass destruction.   However, terrorists do not require superpower funding — instructions for building atomic bombs are available on Internet. Iraq, unable to obtain nuclear technology, continues to defy UN inspection teams mandated under the Gulf War cease-fire sanctions by constructing the "poor man's nuclear bombs," biological and chemical weapons.

The Age Old Question...Palestine

Since the Mid East's boundaries were arbitrarily drawn by the West, Palestine has been a thorn in the Mid East's heart and its barbs have pricked West's behind with oil embargoes and terrorism. Iraq dropping Scud missiles on Israel during the Gulf War should have disclosed that Israel can be a liability in dealing with the Mid Eastern nations and, as we discovered, an expensive liability. It is time for America help solve the Palestinian problem rather that being a part of it.

Since 1991 Yasser Arafat has had to be content with negotiating peace with Israel through a myriad of summit meetings, agreements and treaties to establish a pseudo Palestinian state out of neighborhoods even Israeli goats find distasteful, while flirting with Hamas during the night. The details of the peace agreements were often overshadowed by the news of Israel building Jewish housing in the West Bank, Hamas' indiscriminate bombing of buses, and Joe Sixpack being glued to the television entranced by the O.J. Simpson murder trial or NCAA basketball playoffs.

Unfortunately for the Palestinians, Arafat is negotiating from a position of weakness. He has no military base close to Palestine since Syria chucked him out of Lebanon and he had to move to far off Tunisia because no one in the neighborhood wanted Israeli bombs aimed at the PLO while he was visiting. Not only have the Soviets lost interest in Arafat, he backed the wrong horse during the Gulf War and the Saudis and Kuwaitis cut off his funds. Many Palestinians, Israelis and Arabs — including both those undefinable extremists and moderates — are not satisfied with an American brokered peace, regardless of the claimed noble intentions of United

States Secretaries of State Warren Christopher and Madeleine Albright.[103]

By a twist of fate, the Bush administration turned out to be the best opportunity for a Palestinian peace. For the first time, the United States had credibility and momentum. After the Gulf War, Bush realized that even out of the lying mouth of Saddam Hussein there was a grain of truth in America's double standard. How could America insist Iraq pull out of occupied Kuwait, if it didn't demand that Israel pull out of occupied Palestine? Bush had the right man for the job, Secretary of State James A. Baker III.[104] Baker's cold, Republican businesslike approach was simple: operate from a position of strength. He told Yitzhak Shamir that he was withholding Israel's requested $10 billion loan guarantee until he came to the conference table. As a Texas gentleman, he didn't mention that the United States pumps $3 billion in aid into Israel every year (roughly $600 for each Israeli), nor as Bush's campaign manager for President did he have to allude to the fact that Bush received less than 10% of the Jewish vote in 1988. (Of course, George went wobbly and reinstated the aid when the 1992 election rolled around.)

Baker also took a hard-line with the Palestinians. He told Yasser Arafat he was a rabble-rouser and wasn't wanted at the peace table. This pleased the Israelis, who dislike Arafat, until they discovered that the Palestinians who showed up appeared reasonable in the world's eyes. The Madrid Peace Conference in October 1991 appeared to break ground with the soft-spoken Haider Abdul-Shafi heading the Palestinian delegation, but centuries of hatred and conflict can't be reconciled overnight. In 1992 George Bush lost his reelection bid to Bill Clinton and Yitzhak Shamir was replaced by Yitzhak Rabin.

---

[103] Well-written views of the Palestinian peace process may be found in Edward W. Said's *Peace and its Discontents,* Thomas L. Friedman's *From Beirut to Jerusalem,* and *The New Middle East* by Shimon Peres. Succinct and politically realistic is Avi Shlaim's *War and Peace in the Middle East.*

[104] The former Secretary of State's well-written and frank account of the times may be found in *The Politics of Diplomacy.*

Clinton showed his pro-Israeli position immediately by surrendering his bargaining chip of the $10 billion loan guarantee. The Palestinians received another setback when Clinton reversed United States policy dating back to 1967 by supporting Israel's brazen claim that the lands in East Jerusalem and the West Bank were "disputed" rather than "occupied." The American-sponsored talks were dead. The negotiations shifted to Oslo, Norway. America wasn't involved until it came time for Clinton to come up with some *baksheesh*. The White House peace treaty signing was merely a political photo opportunity, as the Oslo talks left the crucial issues unsettled.

Israel's Prime Minister, Yitzhak Rabin, was gunned down by an Israeli fanatic who believed the peace treaty Rabin made with Arafat was a sell out. Hamas continued to try to disrupt the peace with suicide bombings of Israeli civilians. Arafat couldn't control Hamas and other extremists...assuming he really wanted to stop the bloodshed. Hamas believed Arafat sold the Palestinian people out and still receives funding from Arab nations opposed to the peace process. However, Arafat did manage to pull from under rocks a slimy gang of ex-terrorist gorillas (some may spell it "guerrillas") making up the Palestinian parliament-in-exile called the Palestinian National Council (PNC) to vote on a resolution that Israel has a right to exist. A vocal minority of the delegates, who earned their seat in the PNC by hijacking airplanes and boats and murdering innocent civilians, voted "no," which should tell you what they thought of Arafat's treaty.

Next door, King Hussein of Jordan signed another United States brokered peace treaty with Israel in October 1994. "Brokered peace" is what the press and State Department call the go-between diplomacy. "Brokering" is not the right word. Brokers are supposed to make a commission or get something out of the deal. Instead, Uncle Sap wrote off Jordan's debt of $700 million to sweeten the deal. At the State Department, they said it was cheap since Jordan would never pay the debt anyway.

In May 1996 Warren Christopher announced the brokering of another cease-fire in Lebanon. "Cease-fire" is another misnomer we can't call by its real name, Band-Aid®, because it's the registered trade mark of a bandage for a minor wound, which Webster says provides only "temporary, superficial relief." At the time, Israel was bombing civilians in Lebanon in "Operation Grapes of Wrath" in order to flush out Hezbollah who were firing rockets into the 25 mile strip of Lebanon Israel has occupied since "Operation Peace for Galilee" in 1982.

The complicated cease-fire had to be brokered between Israel, Lebanon and Syria, because Hafez Assad had 35,000 Syrian troops in Lebanon. Assad controlled the borders and allowed Russian Katyushu rockets and other arms to flow from Iran to Hezbollah. Iran should have been invited to the negotiations, but that may have required a tourniquet instead of a Band Aid®. In order to stop the bleeding in Lebanon, Russia must cut off supplying Iran with Katyushu rockets.

In effect, the 1996 cease-fire was merely a rehash of a 1993 agreement for both sides to stop killing civilians and for Hezbollah to quit hiding in the villages behind old men, women and children. The cease-fire also proposed, but didn't guarantee, that Israel and Syria would resume their American brokered peace negotiations or that Israel would return Syria's Golan Heights captured in 1967. If some of these facts sound repetitive, it's because almost the same things happened in 1948, 1967, 1973, 1981, 1982, 1991, etc. As Yogi Berra once opined: "It's déjà vu all over again."

The Band-Aid® had two interesting sidelights — the French sticking their noses into the peace process and claiming "80% of the peace was our idea." It was also brought to light that the Hezbollah terrorists were now legit — the Party of Allah went political and had members sitting in the Lebanese parliament.

What should concern the American taxpayer is the $25 million they contributed to the installation of American Nautilus laser anti-rockets in Israel to shoot down the Katyusha rockets; that the United States is organizing a multi-million dollar international effort to help rebuild Lebanon; and that Yasser Arafat came to Washington with his hand out and left with his pockets jingling. That's not to say that Americans are not willing to shell out a few million bucks to help the peace process and eliminate the suffering of innocent peoples throughout the world. But we should have some semblance of assurance it will bring peace and that it's not just another expensive Band-Aid® applied to a terminal cancer patient or someone about to face a firing squad tomorrow morning.

Let's step back in history.

HISTORY OF LEBANON — 101½ CONTINUED

"In many ways Lebanon is not really a country. It is more like the creation of a committee that operated from a map rather than from any sense of history that produces nationalities and countries. Even the boundaries are artificial, and virtually all the elements of instability are contained in that small, narrow, and unhappy land." Casper Weinberger, *Fighting for Peace.* (We learned that back in Chapter 5.)

Weinberger's astute description of Lebanon can also be applied to the other Arab nations that evolved out of the League of Nations mandates in 1920. At the same time the congenitally arrogant French were creating *le Grand Libon,* the British were drawing lines in the sand and concocting Iraq, with its hodgepodge of Kurds and Arab Shiites and Sunni Muslims. As an afterthought to take care of Abdullah Hussein, Churchill stole a little sand from here and there and conceived Jordan.

Everyone is still looking for the missing pieces from the jigsaw puzzle the United Nations made out of Palestine, which was organized to "maintain international peace and security." Another big loser was the heart of the Ottoman Empire called "Greater Syria," which requires another brief peek back into history.

HISTORY OF SYRIA — 100 ½

Greater Syria under the Ottomans extended into to what is now Iraq, Jordan, Lebanon, Palestine and southern Turkey. A glance at the map discloses that Lebanon was whittled out of Greater Syria's coastline and the rich Bekaa Valley (the hashish capital of the Mid East). In 1939 Turkey was given the coastal district of Alexandretta (now Iskenderun) as a *pot-de-vin* by the French to convince Turkey to side with the Allies, which Syria still claims the French had no right to give away.

Syria's 14 million population is made up of 68% Sunni, 12% Alawite, 10% Christian and a handful of Druze and Kurds. President Hafez Assad runs Syria...exclamation mark! He has also run Lebanon since he tossed Arafat's PLO out in 1981. Assad prefers to say he took over Syria in 1970 during the "Corrective Movement" rather than a coup. As a member of the minority Alawite sect, many said he was not a true Muslim and qualified to rule. Assad settled the question by finding an *alim* (Islamic scholar) to issue a *fatwa* declaring Alawites were Shiite Muslims.

In 1982 after acts of terrorism against his regime by the fundamentalist Muslim Brotherhood, Assad sent an army of 100,000 to Hama, Syria's fourth largest city and the Brotherhood's center. Between 10,000 and 30,000 residents of all ages and sexes massacred. Only archaeologists will ever determine the true number because the army bulldozed the city into rubble and buildings were constructed over mass grave-yards. There have been few vocal complaints by the fundamentalists since the example set in Hama. After barring polit-ical parties opposed to his Ba'ath socialists in 1990, Assad was reelected the following year with 99.98% of the vote accord-ing to official sources appointed by...guess who?

Since the fall of the Soviet Union, Syria has been without superpower aid. In April 1996 the State Department announced Syria was still one of the bad guys on its Terrorist List, making it ineligible for American aid.[105] Thus, if the United States wants to broker a peace between Israel and

---

[105] The State Department Terrorist List of nations supporting ter-rorism includes the dirty seven: Iran, Iraq, Libya, Sudan, Syria, North Korea and Cuba. (Cuba, no longer a real terrorist threat, continues to be guilty of smuggling superb cigars into the United States.)

Syria, it means a little *baksheesh* for Syria and scratching it from the list. Christopher hinted that Syria wasn't as bad as the other culprits — it hasn't personally tossed any bombs recently, but only guilty of granting sanctuary to Hezbollah and the Kurdish Workers Party.[106] Hence, chances are that adjustments to the Terrorist List will be in order...if, and it's a big *IF,* Israel feels safe enough to hand back the Syria's Golan Heights to Assad. This may bring up another list that gives Israel and the United States migraines — the Mid East nations possessing intermediate-range ballistic missiles: Iran, Iraq, Libya and Syria. (Egypt, Saudi Arabia and Israel also have IRBM's, but they're our pals.)

No one knows who will replace President Assad, who is sixty-eight and has a history of diabetes and heart disease. The 12% Alawite minority controlling the government and military were regarded as low-life peasants and infidels by the Sunni majority and everyone else until they gained power by *supporting* the *French* after World War I against the Sunnis. The betting is there is a lot of Sunnis who still think the Alawites are scum bags, and Las Vegas bookmakers say the odds are 8 to 5 that blood will flow before Assad is replaced in another "Corrective Movement."

---

[106] If the United Nations was true to its charter proclaiming "fundamental human rights... dignity and worth of the human person," it would find land for the Kurds spread through Turkey, Iraq, Iran and Syria as a result of the big carve up of 1920. The Kurds would be happy to take any parcel of worthless land no one else wants (doesn't have oil) to stop being starved and slaughtered and go back to fighting among themselves.

HISTORY OF TURKEY — 101 ¼

The Turks remained neutral during World War II until joining the Allies one month before the Germany surrendered. A Turkish diplomat explained that Turkey had learned its lesson after World War I and desired to sit at the peace table, commenting: "We wanted to be on the guest list, not on the menu."

23

OPEC'S POWER & PIZZAZZ — A Cartel it Isn't, But...

OPEC has yet to prove it is a genuine cartel. Economists define a cartel as a combination of producers of a product who join to control its production, sale and price. Linguists point out that cartel also means a written challenge to a duel or an agreement between warring nations to exchange prisoners or rules of conduct for war to assure that civilian facilities, such as telephones, hospitals and brothels, remain open. Some OPEC members meet all three definitions. Iraq and Iran often take prisoners, are hostile and throw down the gauntlet.

Economists debate whether OPEC controlled prices during the "Leapfrog," but generally agree the only real oil shortage and dictating of prices occurred during the OAPEC embargo of the nations supporting Israel and 5% monthly production cutbacks. Otherwise, it was the uncertainty of the industrialized nations whether oil would be available that caused oil prices to shoot skyward. That is the law of supply and demand at its best (or worst) topped off with a dash of panic. OPEC merely took advantage of the market's hysteria.

Unlike Joe Sixpack, who pulls up to his regular Texaco station whenever his gas gauge reads low, Texaco must determine where it's going to get the crude oil far in advance. It takes over a month for a tanker to bring crude oil from the Mid East to Port Arthur, Texas. Texaco must then refine the crude oil into gasoline, put it in a pipeline to a local storage terminal and send a tank truck to Joe's corner Texaco station, which adds another month or so. Texaco is concerned about the replacement cost of the crude oil, which becomes the true market price, because it must always have gasoline at every Texaco station. If Joe's Texaco station runs out of gasoline or Texaco's price is too high, Joe will drive a few blocks to an Exxon or Citgo station. During the 1973 embargo and Iran-Iraq War, when Texaco did not know whether there would be enough crude oil available in the coming months, Texaco and the other oil companies bid up the price. *Unlike the earlier years when the Seven Sisters owned and controlled the production and price of crude oil, they are now buyers.*

OPEC has done a fair to poor job of sustaining higher oil prices by controlling its members' production levels because of their cheating. Today, it can't control the world's oil prices because the majority of today's oil flows from non-OPEC nations. Of the 65 to 68 million barrels per day produced in the world, OPEC provides between 24 to 28 million barrels as the "swing" producer, while the non-OPEC countries produce to capacity.

OPEC has spare production capacity and attempts to satisfy the world's demand non-OPEC nations are unable to produce without overproducing and causing a glut on the market, thereby maintaining higher prices as Adam Smith said they could.[107] When Saudi Arabia unleashed its full production capability in 1985-86, oil prices dropped like a camel turd. Currently, there is a surplus of producing capacity without Iraq's oil because of the UN embargo. When Iraq goes back on full stream, other OPEC members will have to cut production drastically or face a price drop.

The eleven OPEC members will always find it difficult to act as a pure cartel because of their vastly different national interests. That's why they cheat. Several members have no business being in OPEC because they can't afford the discipline OPEC requires to operate effec-

---

[107] According to the Center for Global Energy Studies, there was a worldwide excess capacity of 4.5 million barrels a day (m/b/d) in 1995 based on a demand of 65.2 m/b/d. Most excess production is held by Saudi Arabia: 1.6 m/b/d, Kuwait: 0.75 m/b/d, Venezuela: 0.65 m/b/d, UAE: 0.33 m/b/d, and Iran: 0.30 m/b/d. Iraq's 2.5 m/b/d excess capacity was not counted because of the UN embargo.

tively and they freeload on those with spare producing capacity. The same is true of American independent oil producing companies, who only sell crude oil, and the few large integrated oil companies still owning sufficient crude oil to meet their needs and do not have to purchase crude oil on the open market.

To understand why OPEC will never be a cartel, we have to glance at the statistics in Figures 8 and 9. For those who find economic charts and statistics confusing or boring, forget the title of *Figure 8 — World Oil Reserves & Production.* Imagine it's a fall Monday and you're picking up the newspaper to check the rankings of the top 25 college football teams under a headline: *SAUDIS NO. 1 — U.S. DROPS OUT OF TOP TEN.* During the Olympics, the headline might read: *OPEC SWEEPS TOP SIX SLOTS — BEATS OUT RUSSIA & CHINA...U.S. Barely Noses Out Nigeria to Finish 11th.*

A quick glance discloses that Saudi Arabia is the Michael Jordan of oil reserves, scoring over 25% of the world's total. Ranking second through fifth are its Mid Eastern teammates Iraq, UAE, Kuwait and Iran, giving the big five Mid Easterners 63% of the world's reserves.

Venezuela, although ranking sixth, will never be an OPEC team player because they love their cars and Venezuelans can't jump — Venezuelan crude oil and its national entitlement programs are too heavy.

Figure 9 — OPEC Demographics or, as it might appear on the sports pages, *OPEC TEAM STATS,* show their basic differences and ability to perform. **Saudi Arabia** (Michael Jordan) is backed by **Kuwait** and **UAE,** two quiet "moderates" (Pippen and Kukoc) because of their huge reserves and low populations, which do not require enormous national budgets. The stats show the Arabian peninsula members can afford to hold back oil production to prop up prices, which will also assure that their grandchildren will have oil long after everyone else runs out...And make no mistake about it, the world is running short of oil...It won't run out of oil tomorrow, next year, or in the next decade, but it will run out.

Figure 8
WORLD OIL RESERVES & PRODUCTION

		Reserves Jan. 1, 1996 (million bbl.)	Production 1995 (1,000 b/d)	Reserve Production Ratio yrs.
OPEC (world ranking):				
Saudi Arabia	(1)	258,703	7,867	90.1
Iraq	(2)	100,000	600	
U.A.E.	(3)	98,100	2,194	122.5
Kuwait	(4)	94,000	1,800	143.0
Iran	(5)	88,200	3,654	66.1
Venezuela	(6)	64,477	2,565	68.9
Libya	(9)	29,500	1,370	59.0
Nigeria	(12)	20,828	1,887	30.2
Algeria	(13)	9,200	760	33.2
Indonesia	(17)	5,167	1,329	10.7
Qatar	(25)	3,700	438	23.1
Gabon	(34)	1,340	354	10.4
Neutral Zone	(19)	5,000	400	34.2
NON-OPEC MAJOR PRODUCING NATIONS:				
Russia	(7)	57,000	6,950	22.5
Mexico	(8)	49,775	2,689	50.7
China	(10)	24,000	2,989	22.0
United States	**(11)**	**22,457**	**6,545**	**9.4**
Norway	(14)	8,422	2,720	8.5
India	(15)	5,814	708	22.5
Angola	(16)	5,412	640	23.2
Oman	(18)	5,138	846	16.6
Canada	(19)	4,898	1,798	7.4
United Kingdom	(20)	4,517	2,515	4.9
Malaysia	(21)	4,300	685	17.2
Brazil	(22)	4,200	696	16.5
Yeman	(23)	4,000	345	31.7
Egypt	(24)	3,879	890	11.9
Total Mid East		659,555	18,865	
Total OPEC		778,215	25,228	
Total World		1,007,475	61,445	

Oil & Gas Journal

Figure 9

OPEC DEMOGRAPHICS & OTHER STUFF YOU SHOULD KNOW

Nation	Area (sq. mi.) [size of]	Population (000)	Religion	GDP	Exports (billion)	Imports
Algeria	919,595	28,500	Islam	$ 89.0	$11.7	$ 9.2
Gabon	103,347 [Colo.]	1,200	Christian Majority	5.4	2.3	0.7
Indonesia	741,052	203,600	Islam 87%	571.0	38.2	28.3
Iran	632,457	64,600	Islam	303.0	17.2	23.7
Iraq	167,975 [Calif.]	20,600	Islam	38.0	15.5	6.6
Kuwait	6,880 [N.J.]	1,800	Islam	25.7	10.5	6.0
Libya	678,400	5,200	Islam	32.0	7.7	8.3
Nigeria	356,700	101,200	Islam 50%	95.1	11.9	8.3
Qatar	4,412 [Conn.]	500	Islam	8.8	3.4	1.8
Saudia Arabia	865,000	18,700	Islam	194.0	42.3	26.0
U.A.E	30,000 [Maine]	2,900	Islam	63.8	22.6	18.0
Venezuela	352,144	21,000	Christian	161.0	14.2	11.0

Source: The World Almanac and Book of Facts, 1996

The other two members of the starting five are the "price hawks," not to be confused with the Atlanta Hawks. **Iran**, with a population greater than the rest of the OPEC Mid Easterners combined, needs oil revenues to keep things quiet at home. Also, not being Arabic, make it difficult to be a team player.

Iraq is the Dennis Rodman of the Mid East, always breaking the rules and squandering money on military forces. Iraq's stats don't show its capability to produce a guesstimate of 3 million barrels a day or a reserve-production ratio because the UN suspended it from playing with a trade embargo for kicking Kuwait below the belt.

Venezuela plays in the Western Hemisphere league by its own rules and exceeds OPEC quotas by 400,000 barrels a day. It claims OPEC's quotas are unfair because they are based on total oil production or reserves rather than exports. As Venezuelans love their automobiles as much as Americans, they claim it's unfair. Although its population is comparable to Saudi Arabia and Iraq, its citizens live in a Western democratic-socialist culture and demand more perks from their government. Gasoline is 13¢ a gallon in Venezuela, compared to 33¢ in Saudi Arabia. When the government increased the subsidized gasoline prices in 1989, riots broke out in Caracas leaving several hundred dead. Much of Venezuela's oil is a heavier gravity than Mid East oil, making it less valuable, and costs 80¢ to $1.00 a barrel to produce, compared to between 25¢ to 50¢ a barrel in the Mid East. Socialist-leaning Venezuela, the founder of OPEC because it wanted it to curtail production so it could increase prices, is spending money it doesn't have, causing rampant inflation. Currently, it is making noises that it wants to privatize its new oil exploration areas and obtain foreign investment, which will make it impossible to play under OPEC rules.

Libya and Algeria, ranking ninth and thirteenth respectively in reserves, are OPEC second stringers who lack training and discipline. They have substantial reserves of valuable light sweet crude oil and are a short distance from Europe. However, Libya's Qaddafi is a psychotic price hawk because of his outright greed and fetish to own doomsday weapons, including missiles with chemical warheads. Qaddafi doesn't get along with anyone, including the Muslim Brotherhood in his own country, who rioted in Benghazi in September 1995. There have been several attempts to assassinate Qaddafi, but without luck.

Algeria has been embroiled in what amounts to a civil war since 1988, when the government canceled elections because it looked like the militant Muslim fundamentalists would win. Since Algeria's President was assassinated in 1992, an estimated 60,000 civilians have been massacred or, in the language of the government, were involved in "extrajudicial executions." There is no way of predicting what type of leadership will evolve in Libya and Algeria, but the odds are they will stick with their Arab OPEC team members.

Nigeria, with Africa's largest population, is OPEC's unpolished bench warmer who looked good in the draft but doesn't have what it takes to be a pro. Human rights abuses, corruption and incompetence have crippled the nation. Its economy is dependent on oil for 95% of its exports; thus, it has never been able to cut back its production significantly when OPEC asked. Nigeria has to import petroleum products because of its inability to maintain its refineries and produce gasoline, which sells on the black market for over $2.00 a gallon. If Nigeria left OPEC tomorrow, OPEC wouldn't miss the perennial freeloader and cheater on quotas and prices.

Indonesia's burgeoning population of close to 200 million, relatively low seventeenth world reserve ranking and high rate of production make it a likely candidate to retire from the OPEC squad in the next few years. It is also facing rampant inflation and monetary turmoil due to economic mismanagement. The island nation plays in the Pacific league and its oil concessions and prices are governed by different rules than the Mid East. Japan has yet to break the Seven Sister's Pacific zone defense and pays more than it should.

Connecticut-size **Qatar,** with a population of 500,000, will have to content itself as OPEC's cheerleader even though it squeaked in at twenty-fifth in oil reserves. Its only function *was* to vote with its "moderate" Arabian peninsula neighbors. Things changed in June 1995 when Amir Khalifa al-Thani turned on his radio one morning and heard that his son, Hamad, had taken over the throne. Daddy moved into a palace next door with his friends in Abu Dhabi, taking $4 billion of Qatar's treasury and swearing that he'll toss his ungrateful offspring out at the first opportunity. Saudi Arabia and Bahrain don't like young Hamad's moves towards democracy and have yet to recognize Hamad as Amir. The Saudis aren't too keen about Hamad being friendly with Iran and Israel either.

Qatar should sit on the sidelines with the other **OAPEC** ranked players with relatively significant oil reserves, such as Oman (eighteenth), Yemen (twenty-third), Egypt (twenty-fourth) and Syria, who just missed being world-ranked at twenty-seventh. All are oil exporters and, with the exception of Egypt, have small populations. OAPEC members don't have to play with OPEC, which requires paying dues and team discipline. They can sit in the stands and cheer, then charge the same high prices the OPEC starting five dunk by withholding production.

Tiny **Gabon,** thirty-fourth in the world's oil reserves, was OPEC's batboy. It had no business in OPEC because of its lack of oil reserves and high rate of production. Gabon derived little or no benefit from the $1.8 million annual dues. It's withdrawal in 1997 was long overdue.

For those who enjoy pouring over statistics, it should be apparent that thirteen of the top twenty-seven are from the Mid East and *twelve are Arab nations, making OAPEC a potent force if oil is used as an economic or military weapon. Although OPEC could not control its non-Arab members in the aftermath of the Yom Kippur War, OAPEC WAS UNITED AGAINST THE SUPPORTERS OF ISRAEL...ESPECIALLY THE UNITED STATES. If Iran had its way, it would take over OAPEC and change its name to the Organization of Islamic Petroleum Exporting Countries.*

Then, there is the instability of the Mid East — what happens if Iran or Iraq starts shooting at Saudi Arabia, Kuwait or the UAE?

GET YOUR HAND HELD CALCULATORS AND FIGURE OUT WHERE AMERICA WILL GET 8 MILLION BARRELS OF OIL A DAY IT MUST IMPORT IF SAUDI ARABIA IS INVADED BY IRAQ OR IRAN OR IF THEY LOB A FEW WELL-PLACED MISSILES AT THE SAUDI OIL TERMINALS BEFORE AMERICAN MARINES CAN GET THERE...AND, DON'T FORGET THAT THERE ARE THE TERRORISTS WHO GIVE NO WARNING...SEVERAL WELL-PLACED BOMBS

*AT SAUDI ARABIA'S RAS TANURA OIL TERMINAL COULD
DESTROY ITS CAPABILITY TO SHIP OVER THREE MILLION BAR-
RELS OF OIL A DAY.*

*On the other side of the coin, what happens if Iran and Iraq even-
tually become peaceful-loving neighbors in the Mid East? Peace, a nebulous term often grounded on being tired of war and its human and monetary costs or simple economics, may permit farsighted leaders in Iraq
and Iran to become "moderates" and combine with Kuwait, Saudi
Arabia and the UAE to form a real cartel and control prices. Thus, there
may be a relationship between the price of Mid East oil and the price of
Mid East peace.*

"THERE ARE LIES, DAMN LIES & STATISTICS,"
Mark Twain observed.

The reserve/production (R/P) ratio in *Figure 8* does not
predict the number of years until a nation will run out of oil,
even though the government and economists continue to turn
out the intimidating figures. The Department of the Interior
made that mistake when it concluded the United States would
run out of oil in precisely nine years and three months in 1919.
The R/P ratio merely states *known reserves* and annual rate of
crude oil production, without taking into account potential discoveries of new oil fields and technological advances which
will allow more oil to be produced. We generally leave 25%
to 50% of the oil in the ground because we don't know how to
get it out or it's too expensive to squeeze more out. Also, if the
price of oil increases, it can make unprofitable oil wells eco-
nomical. Nevertheless, the R/P ratio is a handy guide to the
reserve and depletion of a nation's oil reserves, *assuming their
production does not change,* **if one keeps in mind that global
demand is currently rising at over 2% a year.**

Be aware of the statistical fudge factors: *Reserves* are
estimates of "proven, probable or possible" reserves by people

who don't always tell the truth or don't know any better. "Proved" means technology indicates the oil is in the ground where we say it is; "probable" means we think it's there; and "possible" is wishful thinking — we hope like hell it's down there. This brings to mind the oil patch adage: "You never know if there is oil until you drill." *Crude oil production* figures often do not include condensates and natural gas liquids (NGL — gas that is converted into a liquid when it reaches the surface). *Figure 8* shows United States' production at 6.545 million barrels a day; however, when the NGLs are added, the production totals 8.650 million barrels...That's the good news.

The Bad News

Americans have drilled for oil in every likely and most of the unlikely places in the country, with the exception of the Alaskan wilderness and offshore waters, which have been placed out of bounds for fear of environmental pollution and endangering the caribou's sex life. It is unlikely we will find another oil bonanza like the Alaska North Slope. Our reserves and capability to produce oil have declined steadily since 1970, when we produced 11.3 million barrels a day, to 6.5 million in 1995 as shown in *Figure 8*. OPEC nations such as Saudi Arabia, Iran and Iraq still have potential oil regions to explore. Outside OPEC, Russia and China have immense unexplored areas. Both nations are enigmas. No one really knows their capabilities, so toss their stats in the trash after noting that both have more oil than Uncle Sam. During the past seven years, Russia's production has plummeted from 11 million barrels a day to less than 7 million because of wasteful production practices. Today, you can't get a hotel room in Moscow for the hordes of oilmen clamoring to invest billions of dollars and exploit Russia's potential. Progress is being held up by Russia's morass of regulations, taxes, export quotas and squabbles about how to split the profits, plus a tinge of

corruption from   former KGB officials connected with
Russia's capitalistic oil and gas companies.

Caspian Basin Oil
The former Soviet republics are the latest unknown statistic and hottest spot for the international oil companies. Deal
makers and drilling crews are invading the wilds of new
nations your history professor never heard about: Azerbaijan,
Kazakhstan, Turkmenistan, Tajikiustan and Uzbekistan.
Caspian Basin oil is currently being held hostage to politics.
Russia is demanding a piece of the action, working to keep
American interests out, insisting the landlocked oil flow
through Russian pipelines to the Black Sea and making noises
about dealing with Iran to ensure pipelines do not go through
Turkey. The U.S. State Department recently claimed that the
Caspian Basin holds 150 billion barrels of oil in order to entice
American companies to explore, but the oil companies believe
the figure is only between 30 to 50 billion barrels, which
brings us back to Mark Twain's observation...

Lies
During the 1980s six OPEC nations raised their estimated
reserves between 42% to 197% to boost their OPEC production quotas...fairy tales. Two prominent trade journals disagree
on the 1996 world reserve figures. *World Oil* reports the world
reserves at 1,160 billion barrels and *Oil & Gas Journal* at
1,019 billion. Two respected experts claim both journals were
all wet — there are only 850 billion barrels of conventional oil
left in the world.[108] While we're discussing liars, don't believe
the exaggerated claims oil companies make about their oil
reserves, especially if it's in the form of a smooth talking

---

[108] Colin J. Campbell and Jean H. Laherrére's *The End of Cheap Oil*
in the March 1998 *Scientific American* is an excellent article on determining oil reserves and seeing through fuzzy oil reserve statistics.

Texan trying to sell you stock, notwithstanding the Securities and Exchange Commission's regulations. Another tip: don't believe big round numbers, like Iraq's 100,000 oil reserve in *Figure 8* — it's a guesstimate.

HISTORY OF AZERBAIJAN 99 $^1/_2$

After Muslim Azerbaijan cut off Christian Armenia's gas supplies, it became the *only* country in the world forbidden by U.S. law from receiving humanitarian aid. Clinton may not know anything about oil and foreign policy, but he knows there are more Armenian voters in the United States than there are Azerbaijanis. The law is not merely dumb...it is cruel.

QUIZ: When will the world run out of oil?

Answer: No one knows. The question should be: When will oil become sufficiently scarce so that the world is dependent on Mid East oil? Answer: My guess is that between 2010 and 2015 the world will depend on the Mid East for over 50% of its oil. Next Question: Then what will we do? Answer: I don't know and no one else knows for sure.

Oil substitutes, such as tar sands, oil shale and coal gasification, will be expensive and cause additional pollution. The only United States potential for large oil deposits are in Alaska and offshore, most of which is banned for environmental *concerns* (I didn't say "reasons"). We may recover additional oil from depleted fields, but have yet to develop the technology.

And...don't believe Vice president Al "Ozone" Gore if he tells you otherwise.

24

A PEEK INTO A CRYSTAL BALL —
Sandstorms Keep Clouding the Future

Not even the sleazy Gypsy at a county fair will dare predict the future of Mid East and its oil, but every oilman does. They have to. If they don't, they will go out of business. When Joe Sixpack pulls up to his Texaco station, he wants to fill up his tank and could care less where the super high test came from. However, if the oilmen's crystal ball is cloudy and prices jump up, as they did in the spring of 1996, they didn't have to read tea leaves to know that President Clinton would launch a Federal investigation into price collusion by the oil industry (it was an election year).

Politics play a major role in Mid East and oil. The United Nations Gulf War sanctions embargoed Iraq, cutting off its oil exports in 1990. American oil companies are barred from doing business in Iraq, but that hasn't stopped the French, Russians and Japanese from making deals. United States sanctions levied in 1986 bar American companies from operating in Libya. The mild sanctions imposed by the UN in 1992 for Libya's involvement in the bombing of a Pan American 747 over Lockerbie, Scotland, in which 259 passengers and 11 people on the ground were killed, did not stem the flow of oil to France, Germany and Italy dependent on Libyan oil. In April 1995 Clinton suspended trade with Iran for funding terrorism and its plans to obtain nuclear materials from Russia, blocking American companies from doing business in Iran. The Russian-Iranian nuclear deal was still in the works in 1996 after Clinton went to Russia to bolster President Boris "The Boozer" Yeltsin's reelection campaign.

The unlikely dense duo of Senators Alphonse D'Amato (R. N.Y.) and Ted Kennedy (D. Mass.) sponsored legislation imposing sanctions on foreign oil companies doing business with Iran and Libya. The so-called "secondary boycott" threat pissed off America's major trading partners, Britain, Germany, France, Japan, Canada and Mexico. The short-sighted Congress will look even more foolish when it attempts to enforce the secondary boycotts.

Thus, unless the political winds blow the shifting sands from another direction or Iran, Iraq and Libya change coaches, American companies will not be allowed to play in their oil fields.

QUIZ: Do sanctions such as oil embargoes work?

Answer: It's a superb topic for intellectuals to discuss at a cocktail party. If America's purpose is to wreck the economy of a small nation and make its poor even more destitute under a repressive regime, the answer is that it works. If our objective is to force a nation to become democratic through the overthrow of a dictator, the answer is "hell no." Saddam Hussein is a good example. Fidel Castro's Cuba, which has been under an embargo since 1962, proves it doesn't work. More often, sanctions merely reinforce the people's dislike of America, particularly if they are starving, and give a dictator someone to blame for his corruption and economic failures. Also, unless all nations join in the embargo, it's bound to make the United States look arrogant, reckless...and *stupid*.

Clinton's forcing American oil companies to cancel contracts to rebuild Iran's oil fields destroyed during the Iraq-Iran War had a negative impact — he wasn't playing international politics with a full deck. It was a slap in the face to President Rafsanjani, who wanted to improve Iranian-American relations (which qualified him as a "moderate"), and closed the door on American access to Iranian oil for many years by handing it over to other nations not likely to let go of the oil after the embargo is lifted. Iran considered Clinton's actions treachery because it involved the termination of a $1 billion contract with Conoco, an American oil company. The cancellation was lobbied for in Washington by Edgar Bronfman, Sr. — a member of the board of directors of DuPont (Conoco's parent company) and heir to Seagrams, the Canadian booze company — and Israel Singer, the secretary general of the World Jewish Congress. How could Rafsanjani have talked to Clinton about halting terrorism and building nuclear weapons, when the other guys in Iran — those "fundamentalist extremists" — are shouting that the American Satan wants only to destroy Iran's economy and build up Israel?

The only American who is pals with Qaddafi and Saddam Hussein is Nation of Islam leader Louis Farrakhan, who agreed with his buddy, Saddam, that American policy towards Iraq was "wicked" and was causing the "mass murder of the Iraqi people." No one would have noticed his ranting in the desert except that Farrakhan stood out in the crowd wearing a silly bow tie. Their meeting hit the headlines when an overzealous Department of Justice dingbat sent Farrakhan a letter accusing him of being an agent of the Libyan and Iraqi governments.

The press failed to comment whether Farrakhan had paid back the $5 million Qaddafi loaned him in 1985. What the Justice Department was concerned about in 1996 (an election year) was that Qaddafi pledged to give Farrakhan $1 billion for Muslim causes in the United States so Americans would know what a great guy he was and to influence American elections and foreign policy. (Don't hold your breath waiting for the check, Louis.) A year earlier, Qaddafi had suggested America and Libya follow the Arab custom of marriages between offspring to patch up tribal fighting and proposed that his son marry Chelsea, President Clinton's daughter. There was no official White House response to the proposal, but it is rumored that Hillary Clinton said: "I think crazy the brown-neck must have fallen off a camel and hit his head on a rock."

Farrakhan's report of his chat with Saddam Hussein, however, contained a noteworthy ring of Arab culture when he quoted Saddam: "Not to worry about reconciliation among the brothers of the Arab world because they do get in little squabbles but always find a way to work it out." While the Gulf War was more than a little squabble, Saddam's words echo the Islamic adage: *La uman fi'l Islam* — "There are no nations in Islam." Under Islam, there is the *Dar al Islam* ("House of Islam") and the *Dar al Harb* ("House of War"). My Arab friends are fond of repeating the old Arab proverb: *I against my brothers; I and my brothers against my cousins; I and my brothers and cousins against the world.* All Arab conflicts are regarded as temporary and Arab unity as everlasting. Arabs believe they will fight among themselves until an event or common enemy demands unity...Remember the Arab solidarity against Israel and the oil embargo of the United States?

Saudi Arabia, Kuwait and the UAE have the patience to wait and see what happens in Iraq to their fellow Arab, Saddam Hussein. They have the United States military to back them up, although the Al-Saud clan is not anxious to have an American military presence in Saudi Arabia because America is considered to be un-Islamic and pro-Israel by many Saudis. But...do the "moderates" have the time?

Iran is fomenting Shiite discontent in the area. In November 1995 the bombing of the Saudi National Guard headquarters killed seven, including American military personnel. In June 1996 19 American Air Force personnel were killed by a massive terrorist bomb. Even little Bahrain has been subjected to bombings and riots instigated by the Iranian-backed Hezbollah. Egyptian President Hosni Mubarak has warned repeatedly that the Iranian-inspired fundamentalist intrigue is aimed at all moderate Mid East nations. Mubarak's solution was to send a flock of Egyptian fundamentalists to jail.

The political dissent is also internal. Family run nations discourage democracy. If everyone running Saudi Arabia is an al-Saud, in Kuwait they're from the al-Sabah clan, and in Bahrain they're all al-Khalifas, the public has little to say. In 1992 King Fahd established the *majlis al-shura,* a consultive council. Although Fahd appointed the members and the *majlis al-shura* cannot bind the king, at least he didn't appoint any of his relatives.

Saudi Arabia, like America, has economic problems. The good news is that Saudi Arabia did something about it even Newt Gingrich couldn't accomplish. In 1994 it cut spending by 20% and in 1995 by 6%. The massive building and economic programs started during the oil boom of the 1970s and 1980s have been cut back, as was Saudi foreign aid, which ran between 5% to 7% of its gross national product — the highest in the world. However, with neighbors like Iraq and Iran, less populated Saudi Arabia needs an expensive sophisticated military. The Gulf War cost Saudi Arabia $55 billion. In addition to huge defense budgets, the Saudi government provides a morass of government entitlements and subsidies for food and housing and free college education and medical care. As every American politician knows, once you give the voters something cut rate or free, you're not going to get reelected if you take it away...It has become an entitlement.

Another al-Saud problem is that there are too many of them — over 5,000 princes at last count. As it is not proper for a prince to work with his hands, they are involved in government and business and many have their uncallused hands out for *baksheesh.* This presents an opportunity for Saudi groups with high-sounding names, such as the Committee for the Defense of Legitimate Rights (CDLR), to target businessmen and the middle class who see nepotism and corruption undermining the legitimate business community. Accusations were made that Prince Khaled bin Sultan, the general who shared command with General Schwarzkopf during the Gulf War, and his father, the Minister of Defense, made hundreds of millions of dollars on defense purchases via "commissions," but the CDLR failed to back up its claims with proof.[109]

---

[109] A Saudi friend told me such commissions, which American firms are prohibited from paying under the Federal Corrupt Practices Act, are not *baksheesh.* "If the United States government calls them by that nasty word, *bribes,* and says they are illegal," he continued, "many Saudis will do business with the French, who have a more discreet name, *pot-de-vin,* and are always willing to pour a little wine into the pot." American contractors don't like to discuss the problem because they could lose business and incur severe penalties under U.S. law if caught.

ISLAMIC INTERNET

The alleged dirty linen of the Al-Sauds is received in Saudi Arabia by fax and on the Internet from London because of a government controlled press. The CDLR messages are from fundamentalist *ulema* (clerics who claim they know what the Koran really means). Opponents of the Al-Sauds, such as Safar al-Hawali and Salman al-Audah, were booted out of their university jobs in Saudi Arabia and sent to jail before being exiled in London. These are the same guys who are against women driving and the populace having satellite dishes carrying degenerate un-Islamic western ideas into Saudi Arabia that may corrupt the pure. (They are also on CompuServe.)

If the al-Sauds are to survive, they must create productive jobs for their people, particularly the college educated and technocrats, whose unemployment rate is reported at 25%. The average Saudi is not inclined to manual labor, and leaves it to the Pakistanis and Egyptians to dig the ditches, build their homes and do the heavy lifting. Saudi Arabia has to reduce the estimated four million foreigners working in its nation of 18.7 million.[110] That is not an easy task — look at the Mexicans picking lettuce in California, Iranians and Afghans driving cabs in our nation's capital and El Salvadorian bus boys and construction workers in your hometown.

Do Oil and Politics Mix?

Yes, but the mixture makes politics a hell of a lot greasier and dirtier. America and Israel saw how quickly their European and Japanese friends deserted them during the 1973 OAPEC embargo. The Arab

---

[110] Another unreliable statistic is the population of the Mid east nations shown in *Figure 9*. The stated 18.7 million Saudi population is an inflated guesstimate. The CIA and State Department put Saudi Arabia's population between 8 to 10 million.

embargo against the Netherlands failed because the tough-minded Dutch reminded their European fink neighbors that a lot of Arab oil had to come through the Dutch port of Rotterdam. When Carter embargoed Iranian oil imports, after American diplomats where taken hostage at our embassy in Tehran, Japan was in Iran faster than you could say *kamikaze* making a government-to-government deal for the oil that was under contract to American oil companies.

OPEC is also shifting like the desert sands to create a greater share of the profits and jobs for OPEC nations' citizens. The state-owned oil companies went downstream by taking over the refining within their borders and started shipping the more profitable petroleum products in their tankers, skimming profits from Big Oil.[111] They were aware that the four segments of the integrated petroleum industry — production, transportation, refining and marketing — each have profit centers.

Following international oil company practices, OPEC nations acquired refineries outside their borders: Kuwait, Libya and the UAR in Europe; Saudi Arabia in the Far East and United States; and Venezuela in Europe and the United States. In 1995 OPEC owned 9% of Western Europe's refining capacity and supplied 60% of its oil needs. OPEC supplies the United States with over 30% of its oil imports and has substantial refining interests in America. *The Seven Sisters no longer control refining worldwide as noted in Figure 10.*

---

[111] Iran leads OPEC with 32 tankers, totaling 5.5 million dead weight tons, giving it substantial transportation capability. Indonesia, Iraq, Kuwait, Libya, Saudi Arabia, Venezuela and the UAE have fleets of a dozen or more tankers, according to the *1994 OPEC Annual Report.*

Figure 10

WORLD'S LARGEST REFINING COMPANIES

Rank	Company	Capacity (b/d)	Nationality
1	Shell	4,460,277	Neth./Britain
2	Exxon	3,415,655	U.S.
3	Sinopec	2,867,000	China
4	Petroleos de Venezuela	2,500,880	Venezuela
5	Mobil	2,044,525	U.S.
6	British Petroleum	1,812,100	Britain
7	Pemex	1,631,100	Mexico
8	Saudi Arabian Oil Co.	1,521,400	Saudi Arabia
9	Chevron	1,497,150	U.S.
10	Petrobras	1,233,560	Brazil
11	Texaco	1,168,035	U.S.
12	National Iranian Oil Co.	1,167,700	Iran
13	Amoco	994,700	U.S.
14	Agip	984,336	Italy
15	Kuwait Nat'l Pet. Co.	933,900	Kuwait
16	Total	920,759	France
17	Pertamina	804,745	Indonesia
18	Indemitsu Kosan	785,650	Japan
19	Conoco	723,290	U.S.
20	Sun	692,000	U.S.

Source: Oil & Gas Journal (12/18/95)

Libya's 230,000 barrels a day refining capacity in Western Europe plus the one million barrels a day of oil it ships to Europe binds the Europeans to Qaddafi and makes an oil embargo difficult to enforce even if Libya conducts terrorist activities in Europe.

Kuwait is a major petroleum investor in Europe.[112] In addition to refining and petrochemical plants, Kuwait acquired Gulf Oil's gasoline stations and built others, and presently operates thousands of stations in Europe under the *Q-8 logo*. When Prime Minister Maggie Thatcher decided to privatize British Petroleum in 1987 by selling the government's stock, Kuwait picked up 22% of BP's shares. The British didn't think it was cricket for Kuwait, which had nationalized BP in 1975 and paid BP about 2% what the British thought it was worth, to own such a big hunk of the government's former oil company and passed a law limiting Kuwait's interest to 10%. Kuwait, figuring the big gas guzzlers the wealthy Westerners and sheiks drive were a good investment, purchased 17% of Daimler-Benz; hence, don't be surprised at the number of Mercedes you see in the tiny sheikdom.

In 1988 Saudi Arabia acquired a half interest in three of Texaco's refineries in the United States, having a total capacity of 615,000 barrels a day, and now markets gasoline and other petroleum products with Texaco in the 33 eastern states under a joint venture called Star Enterprise — now America's eighth largest refiner. The deal requires Saudi Arabia to supply 600,000 barrels a day to the venture.

---

[112] Kuwait owns Santa Fe International, a minor oil producer in the United States. In 1982 Reagan's Interior Secretary, Jim Watt, attempted to bar Santa Fe from holding oil leases on federal lands under the Mineral Leasing Act of 1920, which bans foreign corporations of nations from holding federal leases that discriminate against American citizens, on the grounds that Kuwait did not permit Americans to own oil leases in Kuwait. The court ruled that Kuwait does not discriminate — it doesn't even let Kuwaiti citizens hold oil leases. Originally, Kuwaiti citizens held 40% of the stock in the Kuwait National Oil Company, but Sheik al-Sabah thought they were making too much money and decreed that the government buy the citizens out... Of course, Sheik al-Sabah is the government.

When Joe Sixpack pulls up to a Citgo station, he probably doesn't realize it is owned by Petroleos de Venezuela (Pdvsa). Venezuelan-owned Citgo is America's ninth largest refiner, with a refining capacity of 544,500 barrels a day and more retail gasoline stations in the United States than any other oil company. In addition, Pdvsa has acquired interests in other American refineries and a marketing agreement with Union Oil of California.

The Seven Sisters must now compete with OPEC and non-OPEC state-owned oil companies, which are investing in downstream operations. Latin America, once known for *Yanqui go home* nationalization, is now opening up again to foreign investment. Recently, Mexico, Brazil, Bolivia, Peru, Venezuela and the former OPEC member, Ecuador, announced steps to privatize their oil industries. Mexico's national oil company, Pemex, purchased a 50% interest in an American refinery. Soviet Russia and several of its former republics are forming oil and natural gas joint ventures with Mid Eastern and Western companies.

Natural gas, oil's sister fuel, is now piped from Russia to Western Europe and gas from Iran will someday enter pipelines to Europe. Eventually, the dependency on an uninterrupted supply of gas from Russia and the Mid East to Europe's homes and factories may bring nations to the bargaining table faster than saber rattling or terrorist's bombs do today.

The capitalistic system is the best hope for easing tensions in the Mid East and improving relations between the Mid East and America. Happy, well-fed people with jobs are less likely to throw bombs than the angry, hungry and jobless. Trade and profits are more apt to shift Mid East national interests than guns and oil embargoes. Oil accounts for 10% of world trade, far more than any other commodity. Thus, it is doubly important that the trading in oil be free and unfettered by government political mandates and price controls.[113] But before we can trade

---

[113] For an economist's view of the issues involved in international petroleum trade, I recommend *Adjusting to Volatile Energy Prices* by Philip K. Verleger, Jr.

IBM computers, Boeing jetliners, Kansas wheat and Coca Cola [114] for Iraqi and Irani oil, we must understand their problems and viewpoints. We begin by having one thing in common — the need for trade to improve and maintain a higher standard of living. We must also remember that they will not be perfect in our eyes, nor will we be perfect in their eyes. New Yorkers still believe that Texans are a bunch of rednecks, and Texans think New Yorkers are funny-talking damnyankees.

Americans have 2% of the world's oil reserves, but consume 26% of the world's oil produced everyday. America's dependency on oil should be considered at every step when the President makes foreign policy decisions concerning the Mid East and the Congress contemplates amending our laws relating to the environment, employment, foreign trade and balance of payments, federal budgets and national security. The alternative may be that one day in the not so distant future Joe Sixpack will have to sit home in the dark and freeze — there may not be enough oil to light and heat his home and his Texaco station has run out of gas. Joe Sixpack may not have a job to go to because there is no oil to manufacture plastics, synthetic rubber, fertilizers, pesticides, or Mrs. Sixpack's rayon sweater or nylons. Hopefully, Joe Sixpack's sons and daughters will not have to go to the Mid East with the 82d Airborne and Marines to get the oil.

One Last Slap at Politicians

Stop tinkering with oil prices and blaming Big Oil for political reasons. As Senator Daniel Patrick Moynihan (D. N.Y.) joked: "It is in the oldest American political tradition that when anything happens (with gasoline prices), you investigate the oil companies."

The increases in gasoline prices in the spring of 1996 engendered the usual partisan puttering around trying to get votes. The repeal of a 4.3¢ gasoline excise tax pushed by the Republicans to lower the cost at the pump was like a fart in a tornado — it wasn't noticed. With the

---

[114] Former President Rafsanjani's son was reported to have held a license for Coca Cola in Iran.

Federal Highway Administration complaining that 61% of our nation's highways are in "fair or poor" condition, they should be spending the excise tax for what it was designed to do — pay for roads — before Americans motorists start disappearing in giant potholes.

President Clinton, failed to live up to his nickname, "Slick Willie," when he pushed the sale of 12 million barrels of oil from the Strategic Petroleum Reserve (SPR) in May 1996 to help deflate gasoline prices. The amount was a drop in the barrel — 16 hours of America's demand — and was sold at a *loss* of over $10.00 a barrel. Apparently, the President is attempting to make up the loss by selling in volume. Worse, we won't know who to blame if we ever have to rely on the SPR during an embargo or war because both the Republicans and Democrats approved the sale in one of their smoke and mirrors budget cutting agreements.

The Republican's Contract with America has merit in its promise to eliminate the Department of Energy, but so far has breached its contract. I recommend that whatever few useful functions are found in the DOE be transferred to the Department of the Interior and a that a separate office be established for the *sole* purpose of liaison with Congress under an Assistant Sub-Secretary for Worrying Over Energy. The office should bear the acronym of ASS/WOE and keep the bureaucrats and Congress happy without screwing up the economy during the next oil crisis.

25

~~CONCLUSION...~~ — TO BE CONTINUED

There is no happy ending...*yet.* The turmoil in the Mid East continues. We will continue to see changes, large and small, good and bad, incidents blown out of proportion by the press and politicians, and tragic events that boggle our mind, such as the terrorists' bomb that killed nineteen American Air Force personnel in Dhahran in June 1996.

The world was startled when Saddam Hussein's two sons-in-law and daughters defected from Iraq. When they returned the men were murdered and Saddam's daughters were given a divorce faster than a Nevada quickie before their ex-husband's bodies were cold. Saddam learned he couldn't even trust his relatives he put in charge Iraq's chemical and nuclear weapons program from blabbing that he still had a few dirty tricks up his sleeve.

Americans thought Saddam showed compassion by releasing two American bozos working for a defense contractor in Kuwait, who were sentenced to eight years in prison for crossing the border into Iraq. Saddam probably figured they were too stupid to see the sixty-foot-wide ditch Kuwait was building to keep Iraqi tanks from making a wrong turn and ending up in Kuwait City.

In 1996 Saddam sent Iraqi forces north into the "safe haven" U.S. jets were supposedly protecting, to eliminate his Kurdish enemies. After executing an estimated 100 Kurds supporting the CIA effort to overthrow the dictator, President Clinton ordered 44 Cruise missiles fired at military targets in the south as punishment to the bad boy. Clinton could have taken sterner measures, such as sending Saddam to bed without any supper. The coalition had disappeared into the sands and the neighbors,

Syria, Turkey and Iran, never did like the creation of a semi-autonomous Kurdish area next door. *Until the Kurds are given some semblance of autonomy, there can be no peace in the Mid East.*

Saddam thrives on crisis and thumbs his nose at UN sanctions before screaming mobs in Baghdad — the same pathetic rabble that crawled on their bellies to surrender to cameramen during Desert Storm — swearing they will die for their leader.

In June 1996 and again in October 1997, we learned that Saddam still had something to hide when he barred the United Nations from inspecting biological and chemical warfare sites and tossed out the Americans arms inspectors. The UN issued resolutions demanding that the inspectors be allowed to continue or the seven-year old sanctions would not be lifted and Iraq would suffer "serious consequences." Whatever that meant, it was plain it didn't mean the UN was going to bomb the hell out of Iraq. Smelling a chance to make a deal, Saddam upped the ante by declaring 360 "palaces and official residences" off limits to UN inspectors, particularly the American members of the UN inspection team he claimed were CIA spies. Clinton, suspecting Saddam didn't really have 360 palaces and residences, sent two carrier groups and thousands of troops to the Mid East at a cost to American taxpayers he's afraid to admit; however, the Saudis and many Coalition friends said: "Don't land the troops in my backyard."

A settlement was reached before Saddam's and Clinton's bluffs were called by UN Secretary General Kofi Annan, which has more loopholes than a crocheted bedspread. Realistically, no one believed Saddam Hussein; however, in a side deal, he managed to increase the amount of oil-for humanitarian aid he can sell every six months from $2 billion to $5.2 billion.

Don't expect Saddam to disappear from power unless someone close to the madman assassinates him. The Clinton administration's fall-back position is to pump millions of dollars into *seventy-three* (not a misprint...73) opposition groups outside Iraq in an exercise in bureaucratic folly, which includes seminars in office administration, accounting, desk top publishing and...in case they run out of money... grant proposal writing. It makes no difference that several "groups" include *only one person* (which flies in face of Webster's definition of a group) or a few *don't have any Iraqi members.* Major groups expecting handouts (they have already received $100 million) are the Iraqi National Congress, which is penetrated by Saddam's spies, and the two Kurdish factions who can't get along with each other — the Kurdish Democratic Party (currently allied with Saddam) and the Popular Union of Kurdistan (controlled by Iran). In case the other 72 can't overthrow the tyrant, a group called "Indict" will receive $5 million to have Saddam tried for war crimes.

Americans must learn not to expect many major improvements in the near future...and expect setbacks. The shifting desert political sands will continue to drift and change. In December 1975, during the Cold War and high oil prices, the United Nations General Assembly voted 72 to 35, with 35 abstentions, in favor of resolution proclaiming "Zionism is a form of racism and discrimination." Following the Gulf War in December 1991, the General Assembly voted to repeal the resolution by a vote of 111 to 12, with 13 abstaining.

HAPPY 50TH BIRTHDAY ISRAEL

1998 witnessed Israel's 50th birthday, which Palestinians call *al-nakba* — the disaster. Israel is no longer a tiny desert *home* for the Jews, but a nation of 5.8 million people, including 800,000 Arabs. Like the United States, it is a nation founded by immigrants. Like the United States, when we stole the land from the Indians, Israel stole the land of the Arabs.

Israel's European Jews are known as Ashkenazi and Middle Eastern Jews are called Sephardi. The Mid East Jews immigrated from Morocco (268,000), Iraq (130,000), Yemen (66,000), Tunisia 53,000) and Ethiopia (50,000). As in America, the black Ethiopians claim they are discriminated against in Israel. America added 80,000 immigrants. The largest block of immigrants came from the Former Soviet Union (925,000). Poland's (340,000) and Romania's (315,000) immigrants far surpassed Germany's (70,000).

Religious divisions are broken down: 82.5% Jewish, Muslim 14.6%, Christian 3.2% and other, including Druze, 1.7%. **(Yes, there are Arab Christians.)**

The Israeli parliament, the Knesset, is one of the most diverse and contentious in the world, with 13 political parties represented. Prime Minister Binyamin Netanyahu's right-wing Likud party holds only 23 seats out of 120, compared to the "moderate" (a relative term) Labor party's 34. The parties generally represent other than social and economic interests (Russian immigrants, ultra-orthodox). Netanyahu put together a weird coalition of right-wingers opposed to Shimon Peres' peace talks with Yasser Arafat — the principle issue in the 1996 election. Today the knesset includes five from the Hadash party (communist) and four Arabs.

Economically, Israel's per capita income ranks Western Europe's with a $17,400 per capita income. Compared to the United States ($27,600), Britain ($19,600) and Kuwait ($16,200), it is doing well, except for an unemployment rate of 8.2%, *if you don't count the Palestinians.*

Again, Happy birthday, Israel. Remember the pogroms the Jews suffered and do not inflict them on others. At fifty, don't expect birthday gifts (foreign aid) from your Uncle Sam forever.

What happened to America's Gulf War friendships and the coalition against Saddam Hussein? It will be difficult, if not impossible, to build another Mid East coalition against Iraq unless Saddam blunders by committing violence against one of his Arab neighbors or the Israel- Palestinian National Authority peace talks bear fruit. The United States is perceived as bias towards Israel by the Mid Easterners. Until the United States is seen as getting tough...*really tough*...with Israel, it lacks credibility in the Mid East. Questions asked by Mid Easterners are difficult to answer: Why is the United States so concerned about Iraq having weapons of mass destruction? (Israel has them.) Why does the

United States supply arms to Israel, but hesitate to sell arms to Arab nations? (Russia, France and China are our friends and sell us arms.) Israel's support of the United States at the UN against Iraq is tantamount to Yasser Arafat endorsing a candidate at a B'Nai B'Rith convention. Naturally, Israel would like to see Saddam Hussein eliminated, *like most everyone else,* and desires a new Iraq to counterbalance fundamentalist Iran.

"Friendship is to be found between individuals, but between nations interests prevail," Prince Khalid bin Sultan was quoted in Chapter 21. *This explains what happened to the coalition against during the Gulf War...*

UNITED STATES v. IRAQ
Balancing Interests

Saudi Arabia: Our best pal in the Arab Mid East hates Saddam worse than we do, but it's leery about backing the U.S. again for fear we won't finish the job, like we failed to do in 1991. Many of its citizens don't want Westerners sullying the Islamic way or killing Arabs, even Iraqis. The Saudis are used to Saddam's belligerent ranting, but don't believe he will do anything really stupid, like toss bombs, because the U.S. will arrive and save their oil in the nick of time. Unmentioned among oil producing nations, when the sanctions are lifted against Iraq, two million additional barrels of oil a day will flood the market and lower prices.

Kuwait: They have greater reason to hate Iraqis that any other nation, but want to make sure we knock off Saddam next time and Iraq's next leader doesn't want to run over them.

Palestine National Authority: They don't like us because we're pro-Israeli, but Yasser Arafat has to sit at the peace table and hope we are fair. They will root for Iraq, but remain neutral the next time we face Saddam, although they may demonstrate and toss a bomb or two in Israel.

Jordan: King Hussein is up to his ass in Palestinian refugees and will remain neutral as he did in 1991. Iraq is a major market for the few Jordanian industries and Jordan can buy cheap oil trucked in violation of UN sanctions.

Syria: We can't ask for Hafez Assad's help unless we promise to help get back the Golan Heights Israel stole from Syria in 1967. Since Turkey and Israel have been conducting joint military exercises, Assad mistakenly believes that Iraq will be its ally.

Turkey: They didn't join the coalition in 1991, but *may* permit the use of their air bases again. The Turks want Iraq to help hold down the crazy Kurds and continue to sell them cheap oil in violation of the UN sanctions.

Iran: They don't like Iraq or the U.S. and love paying the peacenik role by voting against Iraq. They will probably step up their support of the Kurds against Iraq regardless of what happens.

Egypt: "Lukewarm" describes the second largest recipient of U.S. aid's support for military action against Iraq. Hosni Mubarak insists the U.S. settle the Palestinian problem and pressure Israel to sign the treaties banning nuclear, chemical and biological weapons. He asked: If Iraq cannot possess weapons of mass destruction, why doesn't America make Israel surrender their chemical weapons? Even Iran has signed the treaties. By the way, Mubarak added, Egypt will not destroy their chemical weapons until Israel agrees to destroys theirs.

Can there be peace in Palestine? Peace means more than freedom from war. In Palestine, it means more than a peace treaty, there must be a resolution of all the ground issues. There must be a sovereign independent Palestinian state that recognizes the rights of Israel and controls terrorism. Conversely, Israel must recognize Palestine's rights and the rights of Israel's Arab citizens and protect them from oppression. The Holy City of Jerusalem must be shared or there can never be peace. The rights of settlers in the Occupied Territories must be resolved. And last, but not least, the fate of two million Palestinian refugees from the

Diaspora (the word for the dispersion of the Jews after their Babylonian exile) must be settled equitably. If not, terrorism will continue.

SPEAKING OF TERRORISM...

The Organization of the Islamic Conference meeting in Tehran in December 1997, condemned terrorism, but distinguished "terrorism from the struggle of peoples against colonial or alien domination or foreign occupation and their right to self-determination." While they didn't mention any particular nation by name and it's doubtful the aliens they referred to arrived in UFOs, they demanded Israel cease building settlements on Arab land.

There was a promise of peace settlement — a shaky peace, but nevertheless a partial agreement — between Israel and the Palestinians as Shimon Peres and Yasser Arafat struggled through negotiations, knowing neither could satisfy all their countrymen. In June 1996 the Israelis turned Shimon Peres out of office in favor of a hard-liner, Binyamin Netanyahu, who ran on a platform opposed to Peres' peace settlements with Arafat and a Palestinian state in the West Bank and Jerusalem. President Clinton, who had done everything to sway the vote but wear a Peres button during the Israeli election, flushed in defeat at what appeared to be at setback in the Palestinian peace process.

Netanyahu's campaign rhetoric disturbed the Arabs. No one doubts that he must be made to honor Israel's peace commitments or terrorism will continue. At the same time, there can be no peace for Israel without a guarantee of security from outside its borders. UN Resolution 242 passed in 1967 requires "secure and recognized boundaries," not merely the lines drawn in the sand in 1948.

A weak American policy bending to Netanyahu's hard-line will add to the turmoil and Arab distrust of America. Conversely, a soft American policy towards Hezbollah and Hamas will only encourage terrorism against Israel. During the Bush administration, Secretary of State James Baker banned Netanyahu from the State Department building when Netanyahu said in a speech before the Knesset that the United States was "building its policy on a foundation of distortion and lies." The question is whether any American President has the guts to insist that Netanyahu toe the line and demand a fair settlement for the Palestinians.

For the first time since Iraq invaded Kuwait, twenty-one Arab nations held a summit meeting for the purpose of discussing Netanyahu's election. (Iraq wasn't invited.) As expected, the subtle Arabs took a wait and see attitude and merely cautioned Israel not to renege on its peace promises. They also wanted to evaluate the position of the United States. Syria, which has yet to negotiate a peace treaty with Israel and recover its Golan Heights occupied since 1967, was less than subtle in its demands to take a tough stance.

"Extremists" Iran, Iraq and Libya and numerous "militant" Islamic "extremist" groups in "moderate" Mid Eastern countries remain opposed to Israel. "Moderate" Iranian President Rafsanjani's successor, Mohammed Khatemi, received the dubious description "relative moderate." We hear warnings from America's "friends" that the United States has not been an "honest broker" in the peace process. Recently, Egypt's Foreign Minister, Amr Moussa, a credible "moderate" and likely candidate for the Egyptian presidency when "moderate" Hosni Mubarak steps down, complained of America's bias towards Israel, but it was buried on the back pages of most American newspapers.

America's position in the peace talks has a direct relationship to our policy of the "dual containment" of Iraq and Iran. Our Mid East friends cannot accept America's hard line against Muslim nations while we tread softly with Israel. It has undermined Arab belief in America's impartiality and honesty. Our bias is noted not only by the Arab nations, but the entire world. In March 1997, the United States cast two vetoes in the United Nations Security Council calling on Israel to cease building homes in Arab East Jerusalem and the West Bank. When the vote was brought before the United Nations General Assembly on the second resolution, the vote was 130 to 2. Only Israel and the United States voted against the condemnation and invitation for terrorism.

The Palestinians saw a brief glimmer of hope in American policy in May 1998 when Secretary Albright supported Arafat in the negotiations requiring Israel withdraw from 13% of the West Bank contrary to Netanyahu's position of a 9% pullback and set a deadline for Israel's agreement. This brought on a tidal wave of lobbying from American Jewish groups aping Netanyahu's line that "America was dictating" the peace process and placing Israel in jeopardy. Clinton's policies were attacked by Democrat House Minority Leader Richard Gephardt, who was seeking Jewish support in his next bid for the presidency, and Republican House Majority Leader Newt Gingrich, looking for Jewish funds for the Republicans and pandering to the right-wing Christian fundamentalists, who believe in a literal reading of the Bible and anything Israel spouts in their zeal. Americans should ask their Congressmen what was behind the statement by Israeli political commentator, Nahum Barnea, that Netanyahu wasn't concerned about the Clinton's position because: "He [Netanyahu] owns the U.S. Congress."

Many Israeli's favor the 13% pullback as part of the peace process and adhere to the view of Yossi Beilin, the opposition leader, that Netanyahu's claim the surrender of 13% in the West Bank is dangerous is "nonsense." One Israeli politician called it "bopkess"...goat shit. Netanyahu has long been described as the least ideological leader in Likud party history. There is little doubt his tenuous coalition government would fall if he did not allow Jewish settlements in the West Bank to continue to be built or strays from the hard-line of the extreme right-wing in the Knesset.

ISRAELI FUNDAMENTALISTS

Religious fundamentalism is not limited to Islam or Christianity. Israel has the *haredim* — the ultra-religious Jewish right-wing seven percent minority who wield oppressive political and social power. In a recent poll asking Israelis what concerned them most, the *haredim* came in first, topping the Palestinian peace settlement. The weirdos in long black coats and funny hats you see on the TV newscasts don't work or pay taxes, nor do they have to serve in the military in defense of their tiny nation, but spend their time studying and telling everyone else how sacrilegious they are and to toss the Palestinians out of the Promised Land.

The betting by the Arabs and America's European allies is that Clinton will retreat from taking a firm stand against Netanyahu and abandon the Palestinian position...Albright's deadline was sneered at by Netanyahu...1998 is an Election year and Al Gore needs Jewish contributions for his election bid for the presidency in 2000. The powerless Palestinians have been left to negotiate on their own...so look for an increase in terrorism, if Netanyahu stalls the peace process much longer.

One last word on Mid East oil. As this book goes to press, the price of oil has dropped to below pre Gulf War prices. The Western nations are happy. However, Mid East nations dependent on oil revenues for 60% or 90% of their budgets are being crippled. OPEC is passing resolutions to cut back production to increase the price. Nonmembers, including Mexico and Russia, joined in by promising to cut production. Pundits point out many reasons for the drop in the price — economic woes in the Far East, particularly Japan's doldrums, and warm winter in 1997-98 to mention a few. One reason not talked about was the UN foolishly basing Iraq's ceiling on oil for humanitarian aid on revenue without regard to volume. When the original UN resolution was passed, oil was over $25 a barrel, today, it is below $15. It is little wonder that UN Secretary General Kofi Annan, after increasing Iraq's oil for aid from $2 billion to $5.2 billion in order to convince Saddam to allow UN inspectors to search for weapons of mass destruction, stood like a little boy who just made it to the bathroom before he wet his pants. Annan's trickle down his leg allowed the pouring of another million or so cheap barrels of oil into the market without regard to price to the detriment of Saudi Arabia, Kuwait and other Gulf States, Saddam's enemies and America's allies. *Ho-hum...don't worry. Joe Sixpack can fill up at his local Texaco station at the lowest prices in a decade...until the next large price jump or catastrophe America will not be prepared to face.*

Oil is no longer the domestic issue Kissinger thought in 1973 prior to the OAPEC embargo. The Seven Sisters have been replaced by the *Five Mid Eastern Brothers — Saudi Arabia, Iran, Iraq, Kuwait and the UAE and a few of their Arab cousins.* There must be a fair reconciliation with all the Brothers if there is to be a peace in the Mid East and

America can rely on their oil. But that doesn't mean we have to sell out Israel. Israel must be protected, but so must the rights of the Palestinians...*including the plight of the two million Palestinian refugees living outside their homeland everyone seems to have forgotten.*

It's America's turn to walk the tightrope with Saudi Arabia. Hopefully, we won't get dizzy and fall...It's a long way down.

NOTES ON SOURCES

No book on the Mid East and oil is possible without reference to the *Oil & Gas Journal, World Oil, Petroleum Intelligence Weekly, Petroleum Economist* and *Middle East Economic Survey*. These publications were my sources for the myriad of statistics, oil price quotes and to back up the materials in the texts listed in the Bibliography. Not unexpected, the oil prices and number of barrels quoted in these highly respected periodicals did not always jibe. They, too, have different sources.

Not surprising, the books in the *Bibliography* do not always agree as to the facts. Every author has his or her unique theory and bias. No author, including me, can claim total freedom from prejudice. It is often difficult to separate fact from fiction or history from propaganda, which shouldn't come as a shock. History is generally written by the victors. Books on the Mid East are often written by college professors whose knowledge is based on reading someone else's books and a two week tour of the pyramids, reporters who have to write nice dribble or they won't get a visa to get back into nasty country, so-called "experts" formerly with the Department of State or Ralph Nader, and Arabs who have been screwed an amir, the Seven Sisters or the United States Immigration and Naturalization Service.

I urge reading as much on the Mid East and oil as necessary to become knowledgeable on the subject — at least to the point you are able impress the hostess at a cocktail party sufficiently to invite you back to her next soiree to continue your intimate discussion on why American women can't drive a car in Saudi Arabia, but it's okay for them to fly an F-15. Explaining why United States troops can't fly an American flag over the air base in Dhahran, Saudi Arabia, and have to keep out of sight also makes an interesting dinner topic. And men, the Iranian women you see on television all don't wear black sacks and cover their faces because they're ugly. You will find many attractive Iranian women living in Los Angeles and New York...enough said about my unnamed sources.

Browsing through the bookstore should include authors you believe are brilliant (because they agree with you or belong to the same political party) and those you think are screwballs (left wing nuts if you are a Republican). Chances are you will learn something, even if it's only that there are two sides to an issue...assuming there are only two sides...More than likely, you will find that you don't entirely agree with any side. It's doubtful everyone will agree with me on every issue...especially those @%&*# "extremists."

NEWS MEDIA WARNING

WARNING: The news media may be hazardous to the truth. We all know the news media is bias, not only politically but in connection with the Mid East. To continue with a Surgeon General type warning found on cigarette packages: *Reading and watching the eleven o'clock news may cause bias and complicate reasoned thought.*

Keep in mind, newspapers are in business — they sell papers. Television stations make money by selling advertising measured by the number of viewers. To attract customers, they present the news they believe you want to read or catch on TV. They don't think much of your intelligence or their reporters are voyeurs. When Yasser Arafat and President Clinton recently held a new conference on the Palestinian peace process, the press only asked a few (two?) thoughtful questions about the peace talks. Most of the time was devoted to the President's hanky-panky with Monica Lewinsky. *No wonder most Americans don't know what is going on in the Mid East.*

BIBLIOGRAPHY

Adelson, Roger, *Mark Sykes: Portrait of an Amateur.* London: Jonathan Cape, 1975.

Anderson, Jack, and James Boyd, *Fiasco.* New York: Times Books, 1983.

Anderson, Robert O., *Fundamentals of the Petroleum Industry.* Norman: University of Oklahoma Press, 1984.

Baker, James A. III, with Thomas M. Defrank, *The Politics of Diplomacy.* New York: G.P. Putnam's Sons, 1995.

bin Sultan, Khaled, and Patrick Seale, *Desert Warrior.* New York: HarperCollins, 1995.

Blair, John M., *The Control of Oil.* New York: Pantheon, 1976.

Blumay, Carl, and Henry Edwards, *The Dark Side of Power: The Real Armand Hammer.* New York: Simon & Schuster, 1992.

Bromley, Simon, *American Hegemony and World Oil.* University Park: Penn State Univ., 1991.

Brossard, E. G., *Petroleum Politics and Power.* Tulsa: PennWell, 1983.

Cabinet Task Force on Oil Import Control, *The Oil Import Question: A Report on the Relationship of Oil Imports to the National Security.* Washington, D. C.: GPO, 1970.

Campbell, Colin J., and Jean H. Laherrére. The End of Cheap Oil. *Scientific American,* March, 1998, 78-83.

Carter, Jimmy, *The Blood of Abraham.* Boston: Houghton-Mifflin, 1985.

Clark, James A., and Michel T. Halbouty, *The Last Boom.* New York: Random House, 1972.

Cooper, Bryan, ed., *OPEC Oil Report.* London: Petroleum Economist, 1977.

Cooper, Chester L., *The Lion's Last Roar: Suez, 1956.* New York: Harper & Row, 1978.

Crystal, Jill, *Oil and Politics in the Gulf.* Cambridge: Cambridge University Press, 1990.

David, Ron, *Arabs & Israel for Beginners.* New York: Writers and Readers Publishing, 1993.

Donovan, Robert J., *Conflict and Crisis.* New York: Norton, 1977.

Dunnigan, James F., and Austin Bay, *From Shield to Storm.* New York: William Morrow, 1992.

Elm, Mostafa, *Oil, Power, and Principle: Iran's Oil Nationalization and Its Aftermath.* Syracuse: Syracuse University Press, 1992.

Engler, Robert, ed., *America's Energy.* New York: Pantheon Books, 1980.

Farah, Caesar E., *Islam.* Hauppauge: Barron's, 1994.

Fernea, Elizabeth Warnock, and Robert A. Fernea, *The Arab World.* New York: Doubleday, 1997.

Field, Michael, *Inside the Arab World.* Cambridge: Harvard, 1995.

Fox, William F., Jr., *Federal Regulation of Energy.* Colorado Springs: Shepard's/McGraw-Hill, 1983.

Freeberg, Russell W., and Robert Goralski, *Oil & War.* New York: William Morrow, 1987.

Fromkin, David, A Peace to End All Peace: *The Fall of the Ottoman Empire and the Creation of the Modern Middle East.* New York: Avon, 1989.

Gannon, Michael, *Operation Drumbeat.* New York: HarperPerennial, 1991.

Glasner, David, *Politics, Prices and Petroleum.* Cambridge: Ballinger, 1985.

Glubb, John, *A Short History of the Arab Peoples.* New York: Stein and Day, 1969.

Goldschmidt, Arthur, Jr., *A Concise History of the Middle East.* Boulder: Westview Press, 1996.

Gordon, Kermit, ed., *Agenda for the Nation.* Washington: The Brookings Institution, 1968.

Gresh, Alain, and Dominique Vidal, *A to Z of the Middle East.* Trans. Bob Cumming. London: Zed Books, 1990.

Guertin, Donald L., W. Kenneth Davis, and John E. Gray, eds., *U.S. Energy Imperatives for the 1990s.* Lanham: University Press of America, 1992.

Herzog, Chaim, *The Arab-Israeli Wars.* New York: Vintage, 1984.

Hiro, Dilip, *Dictionary of the Middle East.* New York: St. Martin's Press, 1996.

Hodel, Donald P., and Robert Deitz, *Crisis in the Oil Patch.* Washington: Regnery, 1994.

Hourani, Albert, *A History of the Arab Peoples.* New York: Warner, 1991.

Jacoby, Neil H., *Multinational Oil.* New York: Macmillan, 1974.

Jones, Peter Ellis, *Oil: A Practical Guide to the Economics of World Petroleum.* Cambridge: Woodhead-Faulkner, 1988.

Kaplan, Robert D., *The Arabists.* New York: The Free Press,1995.

Kissinger, Henry, *Years of Upheaval.* London: Weidenfeld & Nicolson, 1982.

Kissinger, Henry, *Diplomacy.* New York: Touchstone, 1995.

Klapp, Merrie Gilbert, *The Sovereign Entrepreneur.* Ithaca: Cornell University Press, 1987.

Kohl, Wilfrid L., ed., *After the Oil Price Collapse.* Baltimore: The Johns Hopkins University Press, 1991.

Lacey, Robert, *The Kingdom: Arabia & the House of Saud.* New York: Avon, 1981.

Lamb, David, *The Arabs: Journeys Beyond the Mirage.* New York: Vintage, 1988.

Laqueur, Walter, and Barry Rubin, eds., *The Israel-Arab Reader.* New York: Penguin, 1995.

Lawrence, T.E., *Seven Pillars of Wisdom.* London: Anchor, 1926.

Lenzner, Robert, *The Great Getty.* New York: Signet, 1987.

Lewis, Bernard, *The Middle East.* New York: Scribner, 1996.

Lewis, Bernard, *The Shaping of the Modern Middle East.* Oxford: Oxford University Press, 1964.

Mackey, Sandra, *The Saudis: Inside the Desert Kingdom.* New York: Signet, 1990.

Mansfield, Peter, *A History of the Middle East.* New York: Penguin, 1991.

Mansfield, Peter, *The Arabs.* London: Penguin, 1976.

Marshall, *S.L.A., World War I.* Boston: Houghton Mifflin, 1987.

Melamid, Alexander, *Oil and the Economic Geography of the Middle East and North Africa.* Princeton: Darwin, 1991.

Miller, Judith, *God Has Ninety-Nine Names.* New York: Simon & Schuster, 1996.

Mottahedeh, Roy, *The Mantle of the Prophet: Religion and Politics in Iran.* New York: Penguin, 1987.

Murphy, Robert, *Diplomat Among Warriors.* New York: Doubleday, 1964.

Nekhlek, Issa, *Encyclopedia of the Palestine Problem.* New York: Intercontinental, 1991.

Organization of the Petroleum Exporting Countries, OPEC: *General Information & Chronology.* Vienna: OPEC, 1994.

O'Rourke, P. J., *Give War a Chance.* New York: Atlantic Monthly Press, 1992.

Ovendale, Ritchie, *The Middle East Since 1914.* Essex: Longman, 1992.

Pahlavi, Ashraf, *Time for Truth.* N.p.: In Print Publishing, 1995.

Patai, Raphael, *The Arab Mind,* New York: Scribners, 1973.

Pellegrino, Charles, *Return to Sodom and Gomorrah.* New York: Avon, 1994.

Peres, Shimon, *The New Middle East.* New York: Henry Holt, 1993.

Polk, William R., *The Arab World Today.* Cambridge: Harvard University Press, 1991.

Pratt, Julius W., *A History of United States Foreign Policy.* Englewood Cliffs: Prentice-Hall, 1955.

Randal, Jonathan C., *After Such Knowledge, What Forgiveness?* New York: Farrar, Straus and Giroux, 1997.

Said, Edward W., *Peace and its Discontents.* New York: Vintage Books, 1996.

Salinger, Pierre, and Erik Laurent, *Secret Dossier: The Hidden Agenda Behind the Gulf War.* London: Penguin, 1991.

Sampson, Anthony, *The Seven Sisters.* New York: Bantam, 1991.

Schwarzkopf, H. Norman, *It Doesn't Take A Hero.* New York: Bantam, 1992.

Schweizer, Peter, *Victory. New York: Atlantic Monthly Press, 1994.*

Shirley, Edward, *Know Thine Enemy.* New York: Farrar, Straus and Giroux, 1997.

Shultz, George P., *Turmoil and Triumph.* New York: Scribners, 1993.

Sick, Gary, *October Surprise: America's Hostages in Iran and the Election of Ronald Reagan.* New York: Times Books,1991.

Skeet, Ian, Opec: *Twenty-five Years of Prices and Politics.* Cambridge: Cambridge, University Press 1988.

Slugett, Peter, and Marion Farouk-Slugett, *Guide to the Middle East.* London: Times Books, 1996.

Spillman, Jim, *The Great Treasure Hunt.* Medford: Omega, 1981.

Strohmeyer, John, *Extreme Conditions.* New York: Simon & Schuster, 1993.

Terzian, Pierre, *OPEC: The Inside Story.* Trans. Michael Pallis. London: Zed Books, 1985.

Tippee, Bob, *Where's the Shortage? A Nontechnical Guide to Petroleum Economics.* Tulsa: PennWell, 1993.

Tugwell, Franklin, *The Energy Crisis and the American Political Economy.* Stanford: Stanford University Press, 1988.

Tussing, Arlon R., and Connie C. Barlow, *The Natural Gas Industry.* Cambridge: Ballinger, 1984.

Verleger, Philip K., Jr., *Adjusting to Volatile Energy Prices.* Washington, D. C.: Institute for International Economics, 1993.

Weinberger, Caspar, *Fighting for Peace.* New York: Warner Books, 1990.

Yergin, Daniel, The Prize: *The Epic Quest for Oil, Money & Power.* New York: Simon & Schuster, 1991.

Zillman, Donald N., and Laurence H. Lattman, *Energy Law.* Mineola: Foundation Press, 1983.

82nd Cong., 2nd Sess., Senate Small Business Committee. *The International Petroleum Cartel*, Staff Report of the Federal Trade Commission. 1952.

91st Cong., 1st Sess., Senate Subcommittee on Antitrust and Monopoly. *Hearings on Governmental Intervention in the Market Mechanism.* 1969.

93rd Cong., 2nd Sess., Senate Subcommittee on Multinational Corporations, Committee on Foreign Relations. *Hearings on Multinational Petroleum Corporations and Foreign Policy.* 1974.

JAMES M. DAY

James Day practiced law for thirty-five years, specializing in international and domestic oil & gas and mining matters. As Director of the Department of the Interior's Office of Hearings and Appeals during the turmoil of the Arab oil embargo, he was awarded the Department's Outstanding Service award.

He has taught Oil & Gas Law and the Regulation of Energy at the Washington College of Law, the American University, for fifteen years and was named the Outstanding Adjunct Professor in 1988.

At present he resides in Arlington, Virginia, where he teaches, does petroleum consulting and writes.

FOR OTHER TITLES CALL
800-729-4131